# Improving Collection of Indicators of Criminal Justice System Involvement in Population Health Data Programs

## Proceedings of a Workshop



Jordyn White and Esha Sinha, *Rapporteurs*

Committee on National Statistics

Division of Behavioral and Social Sciences and Education

*The National Academies of*
SCIENCES · ENGINEERING · MEDICINE

THE NATIONAL ACADEMIES PRESS
*Washington, DC*
**www.nap.edu**

THE NATIONAL ACADEMIES PRESS 500 Fifth Street, NW Washington, DC 20001

The project that is the subject of this proceedings was supported by the Department of Health and Human Services through a grant from the National Science Foundation (No. SES-1024012). Any opinions, findings, conclusions, or recommendations expressed in this publication do not necessarily reflect the views of any organization or agency that provided support for the project.

International Standard Book Number-13: 978-0-309-45337-0
International Standard Book Number-10: 0-309-45337-2
Digital Object Identifier: 10.17226/24633

Additional copies of this publication are available for sale from the National Academies Press, 500 Fifth Street, NW, Keck 360, Washington, DC 20001; (800) 624-6242 or (202) 334-3313; http://www.nap.edu.



Printed in the United States of America

Suggested citation: National Academies of Sciences, Engineering, and Medicine. (2017). *Improving Collection of Indicators of Criminal Justice System Involvement in Population Health Data Programs: Proceedings of a Workshop*. Washington, DC: The National Academies Press. doi: 10.17226/24633.

*The National Academies of*
SCIENCES • ENGINEERING • MEDICINE

## STEERING COMMITTEE ON IMPROVING COLLECTION OF INDICATORS OF CRIMINAL JUSTICE SYSTEM INVOLVEMENT IN POPULATION HEALTH DATA PROGRAMS


WENDY MANNING (*Chair*), National Center for Family and Marriage Research, Bowling Green State University

CANDACE KRUTTSCHNITT, Department of Sociology, University of Toronto

JOHN LAUB, Department of Criminology, University of Maryland

RUTH PETERSON, Department of Sociology, Ohio State University

JOSIAH RICH, Center for Prisoner Health and Human Rights, Brown University

MICHAEL SCHOBER, Department of Psychology, The New School for Social Research

EMILY WANG, Department of General Medicine, Yale University

CHRISTOPHER WILDEMAN, College of Human Ecology, Cornell University

JORDYN WHITE, *Project Director*

ESHA SINHA, *Associate Program Officer*

## COMMITTEE ON NATIONAL STATISTICS

# Acknowledgment of Reviewers

This Proceedings of a Workshop has been reviewed in draft form by individuals chosen for their diverse perspectives and technical expertise. The purpose of this independent review is to provide candid and critical comments that will assist the institution in making its published Proceedings of a Workshop as sound as possible and to ensure that the report meets institutional standards for objectivity, evidence, and responsiveness to the charge. The review comments and draft manuscript remain confidential to protect the integrity of the process.

We thank the following individuals for their review of this Proceedings of a Workshop: Carl C. Bell, Department of Psychiatry in the School of Medicine, University of Illinois at Chicago; Michael F. Schober, Department of Psychology, New School for Social Research; and Howard Spivak, Office of Justice Programs, National Institute of Justice, U.S. Department of Justice.

Although the reviewers listed above provided many constructive comments and suggestions, they were not asked to endorse the content of the Proceedings of a Workshop nor did they see the final draft of the Proceedings of a Workshop before its release. The review of this Proceedings of a Workshop was overseen by John V. Pepper, Department of Economics, University of Virginia. He was responsible for making certain that an independent examination of this Proceedings of a Workshop was carried out in accordance with institutional procedures and that all review comments were carefully considered. Responsibility for the final content of this Proceedings of a Workshop rests entirely with the rapporteurs and the institution.

# Contents

# 1

# Introduction

In the U.S. criminal justice system in 2014, an estimated 2.2 million people were incarcerated or under correctional supervision on any given day, and another 4.7 million were under community supervision, such as probation or parole (Kaeble et al., 2015). Among all U.S. adults, 1 in 31 is involved with the criminal justice system, many of them having had recurring encounters. Nearly 3 in 4 prisoners are rearrested within 5 years of release, and more than one-half of jail inmates have been previously incarcerated (Durose et al., 2014).

Although people who are involved with the criminal justice system have gradually gained recognition as being vulnerable to poor health outcomes, including high rates of infectious diseases, behavioral health conditions, and substance abuse comorbidities, the public health implications span far beyond the physical well-being of each individual. Each time a person is removed from his or her home, it affects the family unit and the community in which he or she lives—almost always in a negative way. Disproportionate rates of justice system involvement can exacerbate existing community-wide health disparities, particularly among racial and ethnic minority populations.

The ability to measure the effects of criminal justice involvement and incarceration on health and health disparities has been a challenge, due largely to limited and inconsistent measures on criminal justice involvement and any data on incarceration in health data collections. The presence of a myriad of confounding factors, such as socioeconomic status and childhood disadvantage, also makes it hard to isolate and identify a causal relationship between criminal justice involvement and health. The Bureau of Justice

Statistics collects periodic health data on the people who are incarcerated at any given time, but few national-level surveys have captured criminal justice system involvement for people previously involved in the system or those under community supervision—nor have they collected systematic data on the effects that go beyond the incarcerated individuals themselves.

This proceedings summarizes the presentations and discussions at the Workshop on Improving Collection of Indicators of Criminal Justice System Involvement in Population Health Data Programs, which was held in Washington, D.C., in March 2016. The workshop was organized as part of an effort to assist the Office of the Assistant Secretary for Planning and Evaluation (ASPE) and the Office of the Minority Health (OMH) in the U.S. Department of Health and Human Services in identifying measures of criminal justice involvement that will further their understanding of the socioeconomic determinants of health. The workshop was structured to bring together a range of experts—on incarcerated populations, on indicators of criminal justice involvement, on methods for social measurement, and on the consequences of criminal justice experience on health and life-course events—to facilitate discussion of measures and mechanisms most promising for expanding OMH's data collections in this field.

## WORKSHOP CHARGE AND ORGANIZATION

The charge to the Committee on National Statistics (CNSTAT) of the National Academies of Sciences, Engineering, and Medicine was to organize a workshop to investigate the feasibility of collecting criminal justice experience data with national household-based health surveys (see Box 1-1 for the full charge). To address the charge, CNSTAT formed the Steering Committee on Improving Collection of Indicators of Criminal Justice Involvement in Population Health Data Programs. The main goals of the workshop were to discuss options for collecting data and producing estimates on health indicators of individuals who are or were involved with the criminal justice system, including consideration of the available measures and possible data collection mechanisms.

The workshop was organized in three parts: setting the stage, understanding what should be measured, and identifying best approaches to measurement: see the full agenda in Appendix A.[1] The organization of this proceedings generally follows the workshop agenda.

This proceedings has been prepared by the workshop rapporteurs as a factual summary of what occurred at the workshop. The steering committee's role was limited to planning and convening the workshop. The views

[1]The workshop presentations are available at: http://sites.nationalacademies.org/DBASSE/CNSTAT/DBASSE_170301 [July 2016].

**BOX 1-1**
**Statement of Task**

An ad hoc steering committee will organize and hold a two-day public workshop on adding measures of prior and current criminal justice system involvement to household-based health surveys of the U.S. Department of Health and Human Services (DHHS), such as the National Health Interview Survey, National Health and Nutrition Examination Survey, and others. Justice-involved populations have gradually gained recognition as vulnerable to poor health outcomes, including high rates of infectious diseases and behavioral health conditions. Discussions at the workshop will inform future DHHS selection of measures of involvement with the criminal justice system for potential inclusion in population surveys of health. Workshop topics will likely include the following:

- How do scholars define and measure kinds of criminal justice involvement and their effects on health at the national level, as well as state and local levels?
- If existing measures of criminal justice involvement are available, how well do they work and in what settings or for what types of data collections on population health?
- If measures are not available, what new measures could potentially be developed for inclusion in population health surveys?
- To develop new measures for population-level data collections on how criminal justice involvement affects health, which constructs in each domain need to be measured?
- Do substantial knowledge gaps exist? If so, what are they? Do data exist to support research to fill any remaining substantial knowledge gaps? If not, could such data be generated?

A summary of the presentations and discussions at the workshop will be prepared by a designated rapporteur in accordance with institutional guidelines. Presentations, background materials, commissioned papers, and audio files of the workshop will be posted on the Internet.

contained in the proceedings are those of individual workshop participants and do not necessarily represent the views of all workshop participants, the steering committee, or the National Academies of Sciences, Engineering, and Medicine.

## BACKGROUND AND IMPETUS FOR THE WORKSHOP

To open the workshop, Nadine Gracia (Office of Minority Health) described her agency's goals in exploring how to best measure the effects of incarceration on health. She began with a bit of history. OMH was founded in 1986—in response to a landmark report on the health status of racial and ethnic minority populations—to formulate programs and policies to end the documented health disparities. As OMH celebrates its 30th anniversary, it is an excellent opportunity to reaffirm the commitment to understanding the causes and consequences of disparities in minority health.

Gracia highlighted the health disparities across racial and ethnic minorities and within the majority population and explained how better data and research can help understand points of intervention to reduce such disparities. She acknowledged that significant strides have been made in addressing these disparities, but she also noted that recent events in the criminal justice system have heightened the nation's awareness of the continuing issues. There has also been a heightened sense of urgency, not just in DHHS but across all federal agencies, to better use the available data and to innovate in data collection because of the potential to improve how agencies deal with many issues.

As strides have been made in reducing the health disparities of racial and minority populations, they have highlighted the importance of understanding the socioeconomic determinants of health—criminal justice involvement and incarceration experience being two major factors. Drawing on her experience conducting pediatric work in a juvenile detention center, Gracia explained that physical and mental health needs of people who reenter society after incarceration are significant, and data can play a role in assessing such needs. She noted that the Department of Justice designated the week of April 24, 2016, as National Reentry Week to emphasize the importance of effective reentry strategies for these people, as well as the need to ensure they are given a meaningful second chance in the community.

Gracia noted another event that led to this workshop, a 2012 study and forum on measuring incarceration in household surveys, a collaboration of ASPE, the U.S. Department of Commerce, the U.S. Department of Justice, and the National Center for Marriage and Family Research. This work was in response to the rise in mass incarceration in the United States and the growing number of individuals and families affected directly or indirectly by an incarceration experience. Though the forum and resultant pilot survey, the Survey of Criminal Justice Experience, provided valuable insight, they also showed that critical gaps still remain in terms of understanding the effects of criminal justice involvement on individuals, families, and communities.

Many of those gaps are the result of the disparate measures of history of criminal justice involvement that are collected in the national population health surveys. Thus, although the need for collecting data on criminal justice involvement as part of health data collections has been established, there is no agreement on the scope of the measures to be included. Filling such data gaps is crucial so that programs and policies can be developed to address the many physical health, behavioral health, and social service needs of the populations involved in the criminal justice system, particularly as policy makers consider the reintegration of releasees into their communities and society in general. Given this background, OMH asked the National Academies of Science, Engineering, and Medicine, through their Committee on National Statistics, to organize a workshop to explore the feasibility of collecting enhanced information on criminal justice populations via national household-based health surveys.

## A FIRST-HAND LOOK AT THE CRIMINAL JUSTICE SYSTEM

Glenn E. Martin (JustLeadershipUSA) described his own experience as a convicted offender in the New York State Department of Corrections. He stressed that 65 million Americans have a criminal record, and he quoted Michelle Alexander (from her book, *The New Jim Crow*) saying that more black men are currently behind bars or under the watch of the criminal justice system than were enslaved in 1850.

Martin believes that jails and prisons[2] have become repositories for people who cannot access health care services in the community—a situation that has essentially criminalized poverty, mental health problems, and substance abuse and addiction. Investments in community behavioral health services have declined in the last four decades, and there are currently three times as many people in jails and prisons with mental health issues than in actual mental health treatment facilities. However, even for those without serious health needs, the collateral consequences of imprisonment equate to a lifetime of punishment—the effects of which are felt long after a sentence has been served. Martin used himself as an example of such a long-term effect: in 2015 he was almost prevented from attending an event at the White House to which he had been invited by President Obama because of a 21-year-old conviction (for which he served 6 years in prison).

Martin said the primary goal of the organization he now heads, JustLeadership USA, is to invest in the leadership of people with experiences similar to his and involve them in the discussions about reforming policies and practices. He said the organization believes that people who are closest to the problem are closest to the solution, and the voices of people who have been through the journey are equally important to the dialogue,

[2]In this proceedings, jails and prisons are usually discussed separately because of major differences in nature and length of confinement in them. According to the Bureau of Justice Statistics: "Jails are locally-operated, short term facilities that hold inmates awaiting trial or sentencing or both, and inmates sentenced to a term of less than 1 year, typically misdemeanants. Prisons are long-term facilities run by the state or the federal government and typically hold felons and inmates with sentences of more than 1 year. Definitions may vary by state" (see http://www.bjs.gov/index.cfm?ty=qa&iid=322 [July 2016]).

particularly because of the experiential knowledge they have. Martin said that he believes the United States is in a time of reform, in which the moment has come to reimagine the criminal justice system as well as how it handles the public health of its population.

Martin recounted his childhood and early involvement with the criminal justice system. He was born and grew up mostly in the Bedford-Stuyvesant neighborhood of Brooklyn, New York, the biracial child of immigrant parents from Grenada, and he stood out as different. He witnessed violent crimes in his neighborhood, including being robbed at gun point at the age of 14, and said that he believes that incidents and conditions such as these perpetuate victimization and trauma, which can ultimately lead to more violence. Martin said: "I didn't learn how to pull a gun on someone until someone pulled a gun on me. And that's not to minimize the responsibility of people who engage in criminal behavior, but that is to have a real conversation that moves us away from where we are now about solely focusing on individual responsibility."

Martin's first encounter with the criminal justice system occurred at the age of 16, when he was arrested for shoplifting and charged and treated as an adult offender. (New York is one of two states that automatically charges 16-year-olds as adults.) At the Rikers Island correctional facility, he encountered antiquated, egregiously overcrowded facilities with poor climate control in extreme temperatures, bad plumbing, leaky roofs, limited resources (inmates sharing razor blades, etc.), and lack of safe recreational space (and therefore little to no recreational time). Even at that young age, he realized that the right to health care in jails and prisons was not a guarantee of quality care or even of suitable living conditions. He talked about how in the 1990s New York City opened Rikers Island up to competitive bidding for privatized health care services and how hospitals purposely underbid to obtain the contract, only to provide subpar health care to its 22,000 or so residents—8,000 more than the facility was designed to hold.

Martin recounted his more harrowing personal experiences in the correctional system. He was arrested six times before he ultimately served his 6-year sentence, and several of those arrests also placed him in Rikers, albeit for much shorter stays. Martin witnessed riots and was involved in several fights while he was incarcerated. During his first 48-hour stay, he was stabbed four times with a pen that was fashioned into a knife by another inmate who wanted Martin's leather jacket. The correction officers looked on and laughed, and one warned Martin that if he reported the incident he would not go to court the next day as planned and would be further penalized for trying to go to sick call. As a result, the untreated wound in his neck ended up becoming infected. During another arrest Martin broke his ankle after being pushed down a flight of stairs and physically assaulted by his arresting officer: then he was threatened that, if he wanted the officer to report the injury, the officer would add an assault charge to his file. Martin said that the dissuasion from visiting sick call was commonplace, in part because of the attitudes and threats of the correction officers, but also because the inmates knew the facility would opt for the solution that was most financially beneficial to the franchise and not the patient, such as pulling a tooth with a cavity instead of drilling and fixing it.

Martin also commented on how surprised he was to see so many young, elderly, and physically disabled inmates in prison and how many people resided in mental health units. He told the group a very moving story about an inmate named Arthur, with whom he was bunked during a transitional stay between facilities. Arthur had a mental health condition, and while he was in prison also started using heroin, which Martin said was an incredibly common occurrence; he said he saw more drugs in prison than he ever did growing up in Bed-Stuy. Martin talked about Arthur's unyielding optimism, despite his circumstances, and said that he was a prime example of the type of person who was perhaps ill-served by the justice system. Martin cautioned the workshop participants that one should not move people from being criminalized to being pathologized.

Martin concluded his personal story by describing the obstacles he faced after serving his 6-year sentence, which was 15 years ago. He owed more than $100,000 in debt, fines, court and legal fees, and child support, and in his quest to obtain employment he was turned down almost 50 times in his first month because of his record. Martin was eventually hired as a front desk clerk at a public interest law firm, a job that piqued his interest in learning about and hearing from the people who are directly involved in the criminal justice system. He then went on to work for a long-standing reentry organization that serves thousands of formerly incarcerated people each year.

Martin highlighted the irony that men and women often wind up in the criminal justice system because of poor community health care and resources, only to receive poor health care in that system and then return to the same ill-resourced communities. On average, 600,000 people are released from correctional supervision and return to the community each year, and the effects of this reentry are wide-reaching. It is important to note that the dispersion of this reentry is not evenly spread. In New York, for example, seven communities accounted for 50 percent of the prison population for about four decades, so the impact on those communities was particularly compounded. He closed by urging the participants to push the rhetoric toward providing true opportunities, knowledge, and proper consideration for criminal justice involved populations.

# 2

# Defining Criminal Justice Involvement

There is no current consensus among researchers or policy makers regarding a definition of a person who has been through an incarceration experience. As most of the analysis in this area has been compiled from surveys that collect data for purposes other than capturing information on criminal justice contact or incarceration experience, the questions posed to respondents are not uniform across surveys. Because of these different measures, there is substantial variation in estimates of the variables of interest.

## UNPACKING THE TERM "CRIMINAL JUSTICE INVOLVEMENT"

Chris Wildeman (Cornell University) began by discussing a figure that depicts—in a simplified way—the path of criminal justice contact and all of the potential outcomes that could occur after commission of the crime, flowing through arrest, police contact, charges, bail, conviction, imprisonment, and parole in the "most involved" scenario (see Figure 2-1).

One can use the figure to help in thinking about the intervention points that can be considered for exploring the causal mechanisms at different levels of involvement, Wildeman said. Criminal justice and inmate surveys fall short of capturing all these variations (as they focus on certain subpopulations of the criminal justice community at a certain stage in their experiences), while national non-institutionalized population-based health surveys do not provide information on the prevalence of the health conditions in the incarcerated population. Wildeman noted that it is important to consider not only the variations within the different types of criminal justice contact but also variations across them. He also suggested that although

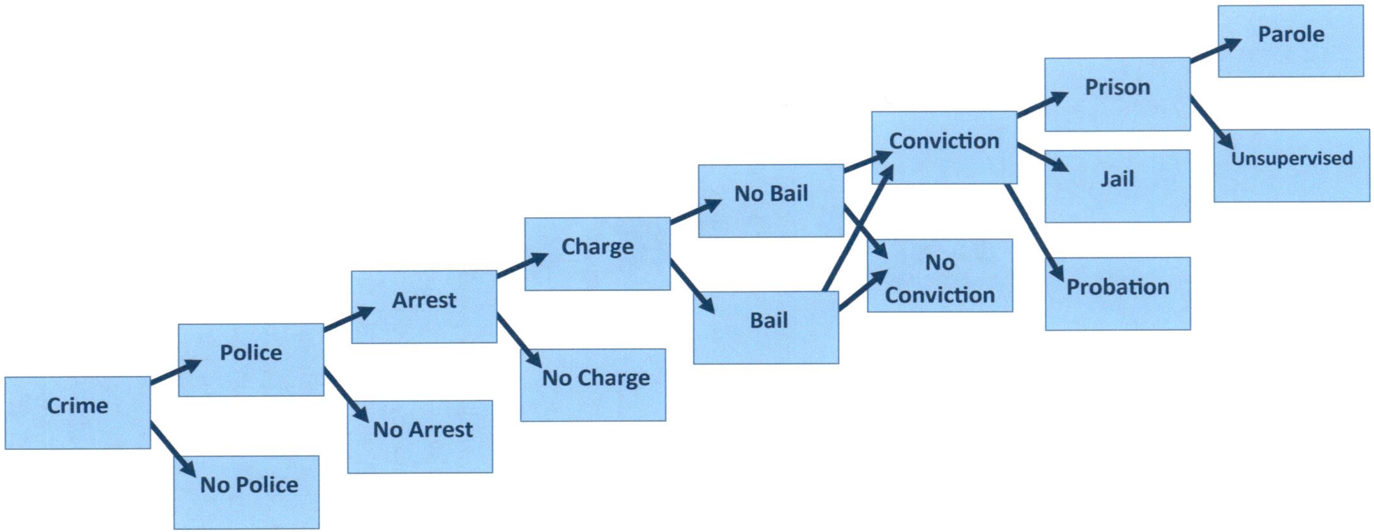


**FIGURE 2-1** A(n insanely) simplified path of criminal justice contact.
SOURCE: Wildeman (2016).

one's first inclination may be to study the causal relationships between criminal justice involvement and health, the criminal justice contact might also serve as a risk indicator. If a person has been arrested, for example, what implications does that lend toward his or her likelihood for certain health outcomes?

Wildeman pointed out that there is limited information available on the health of adults who have experienced the post-conviction stages of criminal justice contact—incarceration, probation, and parole—such as the mortality of inmates and the physical and mental health of inmates. And there is even less information available about people on probation, parole, or no longer under any kind of supervision. In terms of thinking about the connections between criminal justice contact and health, the research community has not gathered many data on the contact points in the "middle" of the figure. And when there are health and criminal justice linkage data for the bookends, they are often collected at a high level and therefore not able to be disaggregated by type of institution or any other degree of contact specificity.

When comparing rates of criminal justice contact within and across subgroups, one would look for disparities between those groups. Wildeman cautioned, however, that the nature of the criminal justice contact questions that are asked could actually influence the relative size of the disparities. He sees this often in his own research, which focuses on racial disparities in the criminal justice system. Crime type, conviction type, and transition probability (likelihood of conviction after arrest) are examples for which one might see varying degrees of racial disparity depending on the measurement variable used: one example is drug conviction rates, which might yield higher racial disparities than homicide conviction rates. The alternative is to look at aggregate totals instead of breakdowns by type, but the tradeoff is loss of the ability to examine the disparities by subgroup.

On the basis of the existing literature and analyses of the effects of criminal justice contact on health, Wildeman said he surmises that the empirical models explain, at best, that 40 percent of the association is causal. The criminal justice mechanisms and the ways in which they affect health differ at every contact level and can also differ within the contact level, as he had noted. The mechanisms driving the relationship between arrest and health, for instance, could be very different from the mechanisms driving the relationship between imprisonment and health. And the context of imprisonment—such as solitary confinement, differing security levels, and distance from family—can all potentially affect the extent to which any mechanism is enabled.

Wildeman noted that measurement becomes even trickier when considering how the contact affects an individual's family and community due to vague measures of criminal justice contact and the omission of specific variables in regression models. He suggested that criminal justice contact questions in national health surveys can be modified to get a better handle on the measure, with the added benefit that the surveys already have information on covariates to test for causal mechanisms of criminal justice contact on health. To achieve those objectives, Wildeman proposed five question topics for inclusion on a nationally representative population health survey:

1. self-reported criminal activity;
2. whether someone has ever been arrested; if yes,
3. ever incarcerated; if yes,
4. type of crime for which incarcerated; and
5. type of facility in which incarcerated.

Wildeman noted that the likelihood of being able to include all five of these questions was slim, but that delving into the type of incarceration could also prove incredibly useful.

John Laub (University of Maryland) asked Wildeman why he placed emphasis on crime and conviction type, especially when 90 percent of cases are resolved by plea bargains, which means that the crime for which a person is punished may not accurately reflect the original act. Wildeman agreed that because of plea bargains, the information that is obtained might create ambiguity for researchers when trying to identify the conviction and its prevalence.

David Cantor (University of Maryland) and Daniel Nagin (Carnegie Mellon University) both commented that learning the nature of the conviction was an important step in determining an appropriate response. Amanda Geller (New York University) agreed that crime type was important, particularly when trying to distinguish between the effects of criminal justice treatment and engagement in criminal justice activity. She also noted that attempting to distinguish the committed crime from the crime to which a person entered a guilty plea will introduce a significant degree of endogeneity.

## MEASURING INCARCERATION IN HOUSEHOLD SURVEYS

Wendy Manning (Bowling Green State University) discussed the key findings from a 2012 forum and study, as well as a subsequent survey that was birthed at the forum.

The National Center for Marriage and Family Research (NCMFR) 2012 forum included the following topics: assessing existing measures of incarceration in household surveys; adapting measures for household surveys; recommendations for future data collection efforts; and linking administrative records data to household surveys. At the NCMFR forum, three surveys that collect data on incarceration experience were compared: the National Survey of Family Growth (NSFG), the 1997 National Longitudinal Survey of Youth, and the National Longitudinal Study of Adolescent to Adult Health. The NCMFR study focused on data collected for male respondents who reported incarceration by age 25. The study panel looked at differences in question content and mode of administration, and weighed the benefits and disadvantages of asking a single question about incarceration experience in comparison with asking multiple questions. The study also pointed to significant variations in reported rates of incarceration, based on the structure of the question(s) about incarceration.

Manning explained that the data disparities among the three surveys motivated the panel to attempt to develop new methods of criminal justice experience measurement that would provide a more accurate picture of how these experiences affect children and families. The panel created an online pilot survey, the Survey of Criminal Justice Experience (SCJE), with items determined by the forum discussions on the most important questions to ask on population surveys for information about criminal justice involvement.

The online survey was administered to 3,200 adults aged 18 to 64, and it had a 62 percent completion rate.[1] The questions were divided into two categories—criminal justice experience and criminal justice supervision—and asked about these phenomena in terms of adult (since age 18) and juvenile (prior to age 18) involvement. The "experience" questions asked about arrest and conviction, and the "supervision" questions asked about probation, jail, and prison. Approximately 20 percent of the respondents reported having had criminal justice experience. Of those, Manning said, about 10 percent pled guilty to or were convicted of a crime. The differences by race, gender, and age range were within expected ranges, with men having higher involvement than women and African Americans having higher involvement than whites and Hispanics.

Manning and her team compared these data to NSFG data for respondents aged 18-44: the rates of criminal justice experience were higher on the NSFG. She said this may be due in part to the way in which the NSFG question is worded, which might encourage respondents to recall any criminal justice experience and not just experience that yielded supervision. Manning suggested to the workshop participants that questions being

---

[1]The survey was administered by GfK, a market research organization, in May 2013 using its online research panel. Panel members were recruited initially through landline telephone using the random-digit dialing (RDD) sampling method. In response to the growing number of cell-phone-only households that are outside the RDD frame, in 2009, GfK switched to the probability-based sampling of addresses from the U.S. Postal Service's Delivery Sequence File (Brown and Manning, 2013).

considered for a survey should be cognitively tested and that researchers should continue to do comparative work to gain a better understanding of what data are being captured across surveys.

Rashida Dorsey (U.S. Department of Health and Human Services) asked what criteria were used in selection of questions for the SCJE and whether or not any cognitive testing had been done. Manning responded that the committee compiled a list of criminal justice experience questions currently asked on national surveys and pared the list on the basis of insight and information obtained and exchanged at the NCMFR forum. The committee then circulated that short list to the researchers who attended the forum and asked for their feedback. Manning reiterated her opinion that any subsequent criminal justice experience question design for a survey should include a round of cognitive testing. Dorsey then asked whether Manning had any history on the pool of existing questions with which the researchers started, to see, for example, if they had been cognitively tested or gone through any prior iterations. Manning said that there was minimal historical information on the origin of the questions on those national surveys.

# 3

# Effects of Criminal Justice Involvement on Individuals, Families, and Communities

There have been many assertions about how the effects of criminal justice involvement may reach beyond the imprisoned individual and affect his or her family, extended family, and the community at large in many different ways. People who are incarcerated endure physical and emotional struggles during the period of imprisonment, and their reentry into the outside world is also full of challenges. It has been argued that the spouses, partners, and children of incarcerated people are likely to face a unique set of financial and emotional challenges that can adversely affect familial relationships. Community dynamics may also play a role: the quality of the environment and the degree of resource accessibility ultimately determine the level of support available to families of incarcerated people and to the people themselves when they reenter society. Although there is some evidence to support these assertions, the research is limited. The workshop session reported on in this chapter focused on identifying key points about the relationship between criminal justice involvement and public health that need to be better understood in order to begin proposing viable solutions.

## EFFECTS ON INCARCERATED ADULTS AND THEIR FAMILIES

John Hagan (Northwestern University) and Holly Foster (Texas A&M University) presented findings from their work on the National Longitudinal Study of Adolescent to Adult Health (Add Health),[1] and on other

[1] Add Health is a longitudinal study of a nationally representative sample of adolescents in the United States who were in grades 7-12 during the 1994-1995 school year. Subjects under-

existing literature about the impacts of paternal and maternal imprisonment on children and families. Hagan began the workshop discussion by noting the significant escalation of the U.S. federal and state prison population over the last century, growing steadily from about 100,000 in 1925 to just over 300,000 in 1980, and then rising rapidly to 1.6 million over the next 30 years. In recent years, however, the incarceration rate has begun to decline. Several states with histories of high incarceration rates, including California, New York, and New Jersey, have seen marked decline (22% or more in 2009-2014), but other states, including Nebraska and Arkansas, are still noting increases.[2] Foster noted that recent studies show that men made up 93 percent of the prison population in 2015. Research shows the influence of male incarceration on people's lack of employment prospects, risk of divorce, diminished political participation, housing insecurity, and homelessness. From a health standpoint, men's incarceration has been linked to depressive symptoms and psychiatric disorders. While women only constituted 7 percent of the incarcerated population, they are a growing minority and an important one because almost two-thirds of them are mothers of minor children. While there is still much to be learned about the female incarceration population, recent data from the Fragile Families Study (see Chapter 7) on maternal incarceration did link the phenomenon to poorer self-rated health.

Foster noted that because one in four women has a family member in prison (Lee et al., 2015), researchers have begun investigating the effects of male incarceration on inmates' female partners. In *Doing Time Together*, Comfort (2008) coined the term "secondary prisonization" to describe the experiences of inmates' female spouses and partners. Turney and colleagues (2012) also investigated how paternal incarceration is linked to diminished mental health and reduced in-kind support for mothers of inmates' children. They found positive association between incarceration and maternal neglect as well as harsh parenting.

Incarceration of either or both parents can have debilitating consequences for children. Foster pointed to research showing that more than one-half of adult prisoners are parents of minor children (Mumola, 2000; Glaze and Maruschak, 2008). Table 3-1 shows a summary of recent research reviewed by Foster and Hagan on paternal and maternal incarceration effects on children, as indicated by children's internalizing and externalizing problems and their social exclusion and marginalization. Foster noted

went four waves of in-home interviews; the most recent was in 2008. Data have been collected from adolescents, as well as from their fellow students in the first wave, school administrators in the first wave, parents in the first wave, and siblings, friends, and romantic partners through multiple data collection components.

[2]The data are from the publication series on prisoners from the Bureau of Justice Statistics.

**TABLE 3-1** Examples of Findings on Maternal and Paternal Influences on Children's Outcomes

| Outcome | Children's Internalizing and Externalizing Symptoms | Children's Social Exclusion and Marginalization |
|---|---|---|
| Maternal Imprisonment | Risk Effects:<br>Huebner and Gustafson (2007)<br>Foster and Hagan (2013)<br>Hagan and Foster (2012a)<br>Lee et al. (2013)<br>Muftic et al. (2016) | Risk Effects:<br>Education: Hagan and Foster (2012a)<br>Foster Care: Swann and Sylvester (2006) |
| | Null and Heterogeneous Effects:<br>Wildeman and Wakefield (2014)<br>Turney and Wildeman (2015)<br>Wildeman and Turney (2014) | Null Effects:<br>Homelessness: Wildeman (2014) |
| Paternal Imprisonment | Risk Effects:<br>Wildeman (2010)<br>Murray and Farrington (2005)<br>Roettger and Swisher (2011)<br>Wildeman and Wakefield (2014)<br>Foster and Hagan (2013)<br>Turney (2014) | Risk Effects:<br>Education: Foster and Hagan (2007)<br>Hagan and Foster (2012b)<br>Turney and Haskins (2014)<br>Homelessness: Wildeman (2014)<br>Powerlessness: Hagan and Foster (2015) |

SOURCE: Foster and Hagan (2016).

evidence in the literature showing that maternal incarceration has adverse or "risk" effects on the manner in which children externalize and internalize symptoms, as well as on their educational success. She said it was also important to note that maternal incarceration was shown to have no distinct effect (null) or a heterogeneous effect (where the incarceration alone could not be identified as the sole factor) in some social and emotional realms for children—namely, behavioral problems and homelessness. A broad range of studies have shown more consistent risk effects of paternal incarceration on children's internalizing and externalizing behavior problems and in the realm of social exclusion, including educational deficits, homelessness, and powerlessness.

Foster and Hagan discussed findings from Add Health regarding the effects of parental incarceration on economic and food security, receipt of medical services, and college access and completion rates for young adults in waves III and IV of the study (Hagan and Foster, 2015). Parental imprisonment was measured on the basis of survey questions that asked respondents whether their biological father or mother has ever served time in jail or prison. Economic or income insecurity was measured by a person's ability to pay for rent or mortgage, telephone, and utility bills; food insecurity was defined by a person's ability to pay for food.

Young adults who were experiencing or who had experienced either paternal or maternal imprisonment were more likely to report income insecurity relative to those adolescents whose parents had never been to prison. The figures were higher for those with incarcerated mothers. Similar results were found in the case of food insecurity. The findings still held after controlling for such covariates as ethnicity, education, marital status, and parental characteristics. Comparing any of the four forms of income insecurity across two time periods—Wave III (2001-2002) and Wave IV (2007-2008)—the authors found that up to 40 percent of the children of mothers who have been incarcerated have problems with one or more of these income insecurities. The rates were higher in Wave IV than they were in Wave III,[3] and the results were consistent across waves in terms of paternal incarceration having a slightly less adverse effect than maternal incarceration.

Another aspect investigated by Hagan and Foster was the linkage of criminal justice system involvement to the receipt of appropriate health care services when needed. There was a negative association between parental incarceration and the receipt of health care services by young adults: fewer young adults received medical services when needed if they had an incarcerated parent, and the effect was stronger for young adults with incarcerated mothers. Multivariate analysis showed that maternal incarceration significantly reduced the odds of receiving medical care when needed, even after controlling for other covariates. However, the analysis also showed that adding health insurance coverage and mitigating economic insecurity reduced the impact of maternal incarceration to non-significance. While both of those factors mediated the effects, the pathway through economic insecurity and marginality was especially pronounced for maternal incarceration. The analysis also showed race disparities in receipt of health services for young adults with incarcerated parents. African Americans were less likely to receive health care when needed than were non-Hispanic whites: 17 percent of the racial disparity was explained by access to health insurance, and 28 percent was explained by economic deprivation.

One of Add Health's unique features is that it contains information on schools attended by children of survey respondents, of which there were upwards of 90 schools. Hagan and Foster said that this information allowed them to study the effects of being in a school in which a large number of the students have had an incarcerated parent—both on children in

---

[3]As the fourth wave of the survey was conducted in 2007-2008, effects of the Great Recession on young adults were also taken into consideration.

that category and on children who did not have an incarcerated parent. In a representative sample of U.S. schools, the rate of maternal imprisonment ranged from 2 percent to 13 percent (with a mean of 3 percent) and the rate of paternal imprisonment ranged from 3 percent to 26 percent (with a mean of 10 percent). Hagan and Foster said they believe that these rates may influence youths' expectations of educational attainment. According to their research, even children with incarcerated parents fare better at schools in which the level of parental imprisonment is near zero. The probability of college graduation for these children is 25 percent; in schools with 20 percent paternal imprisonment, the graduation probability for that same group is cut in half. The graduation probability for children whose fathers were never imprisoned drops from 44 to 25 percent when the paternal imprisonment increases from near zero to 20 percent.

Hagan and Foster also reported on a similar analysis to see the differential effects of attending schools with low and high maternal incarceration rates. Students with incarcerated mothers had a 2 percent chance of graduating college if the school they attended had a maternal incarceration rate near zero. When the school's maternal incarceration rate was close to 10 percent, those students had just a 1 percent chance of graduating college. Students whose mothers were not incarcerated but attended schools with a 10 percent rate of maternal incarceration had a 25 percent likelihood of college graduation.

Hagan and Foster concluded their presentation by emphasizing the societal benefits of government investment in measuring the effects of paternal incarceration on health outcomes. Studies like Add Health allow researchers to investigate health-specific and broader scale outcomes of young adults. Add Health in particular is also able to provide information on racial and ethnic disadvantages, such as how minority youth are more affected by paternal incarceration because they are exposed to the phenomenon more prevalently. The survey has also yielded estimates of effects of paternal and maternal incarceration at community and school levels.

## PUBLIC HEALTH EFFECTS AS A CONSEQUENCE OF INCARCERATION

Christopher Uggen (University of Minnesota) discussed the cumulative effects of criminal punishments. Crime exists, and as such so must punishment in an effort to enforce and enhance public safety. The focus in his work, however, has been to identify instances and areas of excess punishment. Uggen calls the cumulative effects of incarceration and criminal punishment—things like employment barriers, welfare restrictions, or perpetual disenfranchisement—"collateral sanctions." Uggen said that these sanctions are often studied in isolation, which does not give an accurate picture of the way their effects can "pile up" on an individual.

Uggen described the example of Michael, who had several convictions for driving while intoxicated. Michael served his punishments, got sober, and was admitted to college for social work, which in itself was a challenge because of his record. Once he got his degree, however, he had to petition the board to get a social work license because of his record. He still had difficulty finding a job, but did so, moving his family to Minnesota. But when someone on his county's commission took issue with his history, he lost his job—and his family lost their health benefits. Michael did subsequently find employment as a social worker, but it came after having to jump through and over several hoops and hurdles.[4] Uggen also shared the story of Alex, who arrived in the United States illegally as a child but obtained his green card. At age 16 Alex was convicted of statutory rape. He served his sentence, but he did not then register as a sex offender. Alex spent 7 years without criminal incident, but then the U.S. Customs and Immigration Enforcement agency ruled that his failure to register warranted punitive action, and Alex is now facing deportation. Uggen emphasized again that, while some sanctions are merited and essential for public safety, others are capturing a large number of people who may not pose any threat. When one studies the health effects of incarceration, one should also consider the stigmas faced by ex-prisoners and individuals with felony convictions.

Uggen said there are an estimated 2.5 million people currently in prison (Carson, 2015). That total includes 2 percent of the nation's adult males, and 5 percent of its African American males. According to state-level estimates from 2010, 5.1 million U.S. adults have been previously incarcerated—including another 10 percent of the African American male population. He also reminded the workshop participants that while the rates of incarceration have seen recent declines, the number of ex-prisoners continues to grow, so the count now in 2016 is undoubtedly much larger.

Uggen outlined basic statistics on racial composition of ex-prisoners and how they are geographically distributed across the country. Southern and southeastern states have a high concentration of ex-prisoners, as of 2010. African American ex-felons form about 18 percent of the voting age population or about 25 percent of the African American male voting-age population. Some states have high racial disproportionality in criminal justice sysem involvement—notably including California, Illinois, and Indiana—as well as high probation rates and long probation sentences.

[4]As detailed in a recent report on the growth of incarceration in the United States, a growing proportion of U.S. citizens have been denied a range of rights and access to many public benefits, becoming increasingly marginalized in civic and political life (National Research Council, 2014).

Uggen referred to recent research on the effects of incarceration on individuals' mental health. Schnittker and colleagues (2012) found strong evidence for a positive correlation between depressed mood and prison experience. The correlative evidence was less strong for anxiety, panic, and social phobia. From a community standpoint, Uggen and some colleagues tested the hypothesis that states with higher incarceration rates would have strong global effects on disease rates. They found that a percentage point increase in male ex-prisoners led to a 7-17 percent increase in death from HIV/AIDS among non-incarcerated men. Among women, the association ranged from 21 to 27 percent. Their research also showed that the rates were highly contingent on the availability of testing and treatment in a particular area: where the testing and treatment were widely available, having a large number of ex-prisoners in circulation was associated with lower rates of disease than in other areas. Furthermore, the researchers found that the effect of incarceration seemed to change over time: this relationship has declined among men since 2003, which may be due to better detection of diseases in prison and treatment of incarcerated populations.

Uggen laid out the theoretical framework for understanding the spillover effects of ex-prisoners on community:

- Ex-prisoners reentering the community could adversely affect health care utilization in an area through an increase in uncompensated care (care given to those lacking insurance), which could lead to reduction in services and diminished quality. The burden of financing care for the uninsured affects the availability of health care to others, therefore leading to worse community health outcomes in an area. Uggen noted that Schnittker and colleagues (2015) found that as the rate of ex-prisoners rises in a state there are clear effects on health care:
    - Accessibility: residents have less access to care and specialists, reduced physician trust, and less satisfaction with care.
    - Capacity: care received by the uninsured people reduces the number of open beds and appointments.
    - Quality: higher quantity diminishes the quality of ambulatory visits, mammograms, and annual tests.
    - Breadth: insured people, women, adults over age 50, and non-Hispanic whites are most affected.

Uggen ended his presentation by listing some recent policy changes that he believes will continue to reduce collateral consequences for incarcerated individuals, their families, and the community as a whole. He noted that more than 20 states have banned employers from asking questions on criminal history on job applications, which might improve the chances of ex-felons finding employment. Twelve states have opted out of the ban that denies welfare benefits to people with drug felonies. Although only two states allow felons to vote from prison, 19 states restore voting rights after a person's incarceration, parole, and probation are completed. These are significant policy changes that reduce the number of sanctions on ex-prisoners.

## DISCUSSION

John Laub (University of Maryland) asked Hagan and Foster whether wave V of Add Health will include questions on the Affordable Care Act, which would enable researchers to see the benefits of the expansion of health care coverage to previously uninsured populations. Josiah Rich (Medical School, Brown University) suggested that the researchers would want to disaggregate the results by expansion and non-expansion states. Hagan said that the fifth wave of the survey is being conducted and will continue until 2018, which will provide the opportunity to investigate the effects of the Affordable Care Act. On the basis of an analysis done on wave IV of Add Health, he conjectured that the Affordable Care Act would further mediate the effect of maternal incarceration.

Workshop chair Wendy Manning (Bowling Green University) asked the three presenters to comment on remaining data gaps and benefits that can be derived from expansion of available data from other national data sources. Uggen described the difficulties in teasing out state-level and institution-level variation from the data, which would be better suited to produce aggregate statistics. Foster said she would like to broaden the scope of research to incorporate variation in state policies. On this point, Hagan mentioned a study by Beckett and Western (2001) that looked at the effects of the stringency of regimes on welfare support. He believes that such variations over time and across states can be captured by a nationally representative survey such as Add Health.

Ruth Peterson (Ohio State University, emerita) asked Hagan and Foster if the Add Health Study intended to look at children of the original respondents who are now parents themselves, potentially affecting the behavior of their own children in similar ways. Foster said she hopes that the study continues and is able to interview these adult children, enabling them to look at the intergenerational effects of incarceration.

Alexis Bakos (Office of Minority Health) commented on the phenomenon of maternal imprisonment leading to greater economic and food insecurity. She hypothesized that mothers tend to have larger networks (grandmothers, sisters, aunts) and therefore greater social capital than fathers. Bakos also wondered about the interaction effects of paternal and maternal incarceration—whether there were studies on households in which both parents had histories of incarceration. Foster noted the work of Lee and colleagues (2013) at the U.S. Centers for Disease Control and Prevention, whose findings were stronger for fathers than mothers. Hagan referred to the work done by Wakefield and Wildeman (2011) on the use of foster care with maternal incarceration. He added that the research is moving toward looking at aftereffects of foster care and how Medicaid is linked to these arrangements, which is also pushing the cause for having longitudinal data. Candace Kruttschnitt (University of Toronto) pointed out that a mother's addiction habit can be a confounding factor in the relationship between maternal incarceration and food insecurity. Hagan and Foster agreed with Kruttschnitt and said that the next wave of Add Health will allow them to use a fixed-effects analysis to account for this confounding factor.

Daniel Nagin (Carnegie Mellon University) mentioned alternative data sources, such as the Pathways to Desistance Study, which have collected information on health care services received by 1,500 juveniles in Philadelphia, Pennsylvania, and Maricopa County, Arizona (Phoenix). He also highlighted the utility of administrative data by describing his work using court data on convictions and linking it to the Department of Corrections' records to understand the effects of incarceration on recidivism. Nagin added that administrative data are a valuable source of information for studying dose–response relationships (if incarceration is perceived as a dose) and how length of incarceration is related to different outcomes.

In response, Ingrid Binswanger (Kaiser Permanente Colorado and University of Colorado School of Medicine) said that medical records in state and county criminal justice agencies are stored centrally and mostly are not digitized, and she emphasized the need for electronic health records in the correctional health system. She acknowledged that most of these health records are not harmonized, and some of the records are proprietary, which creates data access challenges. The information in those health records can provide prevalence estimates for certain diseases and quality of care received by inmates. Nagin added that developments in the field of text analysis can increase the efficiency of extracting useful information from administrative health records.

Amanda Geller (New York University) added that for her Fragile Families study the researchers were able to get New York State's Division of Criminal Justice Services to match their respondent identifiers to records in the system that had been reported from police departments. That method had a lot of potential but also a lot of pitfalls, such as funding and confidentiality issues, which she will address in her presentation (see Chapter 7). Kathryn Porter (National Center for Health Statistics) recommended the National Survey of Children in Non-Parental Care, conducted by National Center for Health Statistics, as an alternative data source. The survey was a follow-up to the National Survey of Children's Health and looked at children in formal and informal foster care situations (such as kin care). The survey has a module on adverse childhood experiences, which includes parental incarceration.

Michael Massoglia (University of Wisconsin) mentioned a book, *Give Us the Ballot: The Modern Struggle for Voting Rights in America* (Berman, 2015), which argues that 30 states have actually recently enacted more stringent voting restrictions for ex-offenders. He asked Uggen if he is really optimistic about the direction the country is moving or if the gains are negated by the losses. Uggen stated that in terms of re-enfranchisement, he believes the country is moving in the right direction. Voting restriction, however, is a much broader issue, but it definitely includes and is affected by collateral sanctions. There has been work toward change, but with technology the records "stick" with individuals and are harder to overcome.

Chris Wildeman (Cornell University) and Emily Wang (Yale School of Medicine) pointed out that health surveys provide subjective measures on health while administrative records provide information on health indicators of prisoners while they are in prison—not before and not after incarceration. Such data gaps hinder the ability to gather information on the odds and prevalence of diseases for criminal justice involved populations. Wang stressed the need for indicators or variables that simultaneously capture criminal justice involvement and objective measures of health.

# 4

# Criminal Justice Involvement and Health

## INCARCERATION AND HEALTH: CURRENT STATE OF KNOWLEDGE

Michael Massoglia (University of Wisconsin) began his presentation on the current state of knowledge with reference to estimates on the number of incarcerated and ex-incarcerated individuals in the United States. Crudely put, Massoglia said, the same number of people are graduating college as being released from prison: the number of people released from prison each year has fluctuated between 600,000 and 700,000 over the last decade (the vast majority of whom are males) (Guerino et al., 2012); according to the National Center for Education Statistics, 678,000 males received bachelor's degrees in 2008. He noted that there has been a significant expansion in criminal justice involvement in the past several decades. This expansion of the penal system has not occurred in a vacuum or because of a single new law or initiative: it has resulted from a pattern of interconnected laws—state, local, and federal—and it has had effects at all societal levels.

Recent research has studied not only the effects of incarceration on incarcerated adults, but also the health effects on the partners and children of incarcerated individuals and the communities to which they belong. The evidence shows that incarceration generally has negative effects on mental and physical health, for both morbidity and mortality, as well as increasing the risk of anxiety and depression. Those effects on health can be both short term and long term. Massoglia noted that the effects do not differ significantly by race: rather, the effects (including by race) are more often a function of differential exposure, with communities that have higher rates of incarceration being more susceptible to negative health effects.

Massoglia acknowledged that prison conditions can be very harsh (see Chapter 1), but he pointed out that incarcerated people can derive some benefits from the penal system that are comparatively better than what they were exposed to or had access to in their communities, including consistent dwelling spaces, basic sustenance, possible removal or protection from other risks (such as protection from domestic abuse), and access to health care.

Massoglia told the group that identifying a single treatment variable through which to measure criminal justice contact has proven challenging. There is also a question, however, as to whether the selection of one treatment variable over another would have much bearing on understanding the effects of incarceration. He noted the variables that are often considered: whether a person has been convicted of a felony; if that conviction led to incarceration; if so, whether the person spent time in a jail or prison; if applicable, whether federal or state prison; and length of incarceration. Massoglia offered three examples on how greatly the outcome can vary for convicted or incarcerated individuals:

- Convicted felons spend an average of 37.5 months in federal prisons, which involves extended separation from family and community, as well as removal from the labor market.
- About 65 percent of jailed individuals spend one night or less in a jail and may not ultimately even be convicted of a felony.
- It is possible for a person to be convicted of a felony and not spend any time in a prison or jail.

Being arrested has not been a major component of the research on the effects of criminal justice involvement. Since there is no consensus in the current research about which particular measure of contact with the criminal justice system to use, it raises the question of whether having a single, uniform definition or measure can encompass the differential effects that criminal justice contact can have on individuals, families, and communities.[1] Massoglia said that it is important to consider each criminal justice treatment and whether it leads to different health outcomes than other treatments.

Massoglia described measures that are important in identifying the causal relationship between incarceration and health. He approached

[1]Massoglia drew an analogy with how the word "employed" encompasses a number of dimensions that include full-time or part-time status, type of occupation, dual earnings and that, as such, saying one is employed does not provide much insight into the person's specific condition or expected outcome.

the issue from both perspectives: from a health standpoint, longitudinal data can provide the degree of detail needed to identify patterns and problems. Information like help-seeking behavior, lifestyle indicators (e.g., exercise, drug use), and even self-reported illness can prove informative in this regard. From a criminal justice perspective, longitudinal data can also prove very useful, as well as data on type and length of contact and conviction history. Self-reported criminal behavior can provide unique insight and allow researchers to differentiate the effects of being in prison from the effects of leading a risky lifestyle out of prison. Many of these data efforts would require collaboration across multiple agencies at several levels, such as the Department of Justice, the National Institutes of Health, and state and local health and corrections agencies.

Limited data on criminal justice involvement can be gleaned from household surveys such as the Fragile Families Study, the National Longitudinal Study of Adolescent to Adult Health, and the National Longitudinal Survey of Youth. These surveys collect meaningful information, but because the primary focus differs slightly for each, and none of the three is focused primarily on criminal justice involvement, they serve as inconsistent measures of treatment variables for health as it pertains to incarceration. (See Chapter 5 for information on how national household-based population health surveys collect information on incarcerated populations.)

Massoglia outlined the key research questions that he is hopeful can be answered by investments in new survey data collection: causal mechanisms, comparison groups, confounding factors, drug use, global effects, first-order and second-order effects, demographics, and breadth of effects.

**Specific Causal Mechanisms** There is a general consensus in the literature that incarceration matters, and research has provided evidence that suggests that incarcerated individuals are at a health disadvantage. Why, though, does incarceration matter? Is it because an individual has been removed from the community and so lost social support? Is it the exposure to a unique set of stressors and adverse health effects during incarceration? Or is it possibly the stigmatization people face as they attempt to rejoin a community after release?

**Comparison Group** What is the baseline for measurement? Should individuals be treated as their own control, or should researchers make intergroup comparisons of people with similar levels of criminal justice contact? By taking the latter approach, incarcerated individuals could be compared with: ex-incarcerated individuals; those in jails but not in prisons; those who have been jailed but not convicted; and those who have only been arrested. It is straightforward to get estimates or counts of incarcerated populations from administrative and survey data. But for other populations, such as people formerly involved in the criminal justice system, getting accurate data would be considerably more challenging.

**Confounding Factors** How great a role does institutional variation among incarceration facilities play in determining the health effects on individuals? Most health measures only ask if a person was or was not incarcerated. If one views incarceration as a black-or-white effect, the length of the term may not matter much. However, if one considers such possibilities as whether or not the effects are exacerbated by the quality of facilities, the type of confinement (e.g., solitary confinement or in the general prison population), or whether multiple incarcerations yield duplicative effects, the duration of stay would matter.

**Drug Use** Drug use is an important comorbidity/confounding factor, albeit one that is very difficult to measure in this context. Incarceration may prevent access to some drugs but enable exposure to others. However, it is clear that drug use has a detrimental effect on health, and it is an important factor to consider for this population. As essential as longitudinal data can be, they do not always shed light on the effects of drug use because the health problems associated with drug use may have a long dormant period.

**Global Effects** The field is still in the early stages of understanding the relationship between incarceration and health. As mass incarceration is a relatively new phenomenon, it is still unknown whether the health effects of incarceration are likely to become stronger when the incarcerated and formerly incarcerated populations age. If the health effects become stronger, it will have implications for government-funded health programs, such as Medicare and parts of the Affordable Care Act.

**First-Order and Second-Order Effects** As noted above, being incarcerated exposes a person to a regular regime of physical exercise, activity, and food—access to certain daily life activities that may have been lacking for those who come from the lower income strata of society. Are positive effects stronger for certain demographic groups and do those effects become negative over time? Do they ever outweigh the effects of exposure to the negative stressors of incarceration?

**Demographics (Gender, Race, and Age)** Female offenders are the fastest growing group in the prison system, yet the field lacks sufficient data on the effects of incarceration on their health outcomes. Race is another important factor: Hispanics and non-black minorities are particularly understudied populations. In addition, the age at which an individual experiences criminal justice involvement can be another mediating factor when it comes to measuring the relationship between incarceration and health.

**Breadth of Effects** How does the incarceration experience influence the behavior and mental health of incarcerated and ex-incarcerated persons? Massoglia said that some of the above questions can be answered using preliminary and aggregate-level data, but for some questions longitudinal and individual-level data would be necessary. As a starting point for all concerns, there is a great need for population counts. The count of currently incarcerated individuals is readily attainable, but as noted above the numbers of ex-prisoners and ex-felons in the United States are considerably harder to attain. Those numbers are key, however, to considering mortality, migration, reincarceration, and other societal effects.

Massoglia stressed the following next steps to reduce the negative impacts of incarceration on health. First, there is a need to identify best practices and intervention points, both for individuals while they are institutionalized and after they are released from prison. Health-related best practices could include health education, access to drug treatment, and perhaps even adding a health care component to post-release monitoring or parole. Socially, programs for prison inmates to help them maintain family bonds and pre-release counseling may have a positive effect on health. Other programs, such as job training and skill development, can help prisoners find stable employment opportunities once they are released. These social and health components can be mutually reinforcing.

## DISCUSSION

Candace Kruttschnitt (University of Toronto) pointed out that many female prisoners have children while they are in prison, and so their needs are separate from those of male prisoners. She suggested that a longitudinal survey could collect information on prenatal care and follow up on children born in prison. Massoglia agreed that female prisoners do have different medical needs, but the research is lacking on how they are different from male prisoners and how health effects differ for them. In that respect, he agreed that longitudinal data would provide an opportunity to understand those differences and figure out intervention points.

Daniel Nagin (Carnegie Mellon University) commented that he does not believe it sufficient to say that each individual could potentially serve as his or her own control in a longitudinal study. Different "treatments" will have different relative effects, based on an individual's life course, therefore causing the baseline for comparison to potentially change. In terms of medical care, specifically, Nagin commented that it is important to remember the context for comparison: Is one comparing the quality of care before prison to the care in prison? Or is one comparing the quality of care in prison to other, better health care offerings? Massoglia acknowledged that it is important to understand the counterfactual—to whom is one comparing the incarcerated population and what method is being used for the comparison?

Given the data limitations in the field, when information is not always available on a control group, Massoglia suggested the analytical approach of using a longitudinal fixed-effects model. John Hagan (Northwestern University) said that the majority of the research work in the field of incarceration and health has been a reapplication or building on datasets of developmental criminology, psychology, and life-course sociology. He referred to the point made by Massoglia about looking more closely at the experiences of subpopulations within the incarcerated community, which would require a different method of sampling and different allocation of research resources. Hagan asked Massoglia if he thought it was time for that kind of new consideration and reallocation. Massoglia answered that he would first paint with broad strokes, looking at basic demographics—race, gender, etc.—and then move to identifying other causal relationships.

Alexis Bakos (Office of Minority Health) asked whether or not the current research on quality of care in prisons and jails has taken into account the possible iatrogenic effects resulting from subpar practices before major reform changes, such as the Affordable Care Act, were implemented. Could one say that in-prison care was better than community care under those conditions? Massoglia responded by saying that because the data on quality of care and the research community's understanding of it are both limited, it is difficult to ascertain.

Josiah Rich (Brown University) commented that the difference between prison facilities in the United States and those in other developed countries, and between prison and civilian medical facilities stateside, is a lack of oversight. While nonprison medical facilities are subject to the Joint Commission and the Accreditation of Hospitals, where deficiencies can have serious fiscal and other consequences, prison medical facilities do not have anything similar. There is a voluntary accreditation program through the National Commission on Correctional Healthcare, but only about 20 percent of facilities (about 500 nationwide) participate, and lack of accreditation has no significant consequences. Information on quality usually surfaces only when it is truly substandard and there is litigation; however, the Prison Litigation Reform Act has substantially limited the use of litigation as a tool to drive improvements and oversight of health care in correctional facilities.

Emily Wang (Yale School of Medicine) noted that, while causality is important, there are crucial questions that are as of yet unanswered about what the effects of having a large incarcerated population have had on society and about how incarceration actually leads to negative outcomes for individuals, their families, and communities. In terms of effects, how does serving a continuous 30-year sentence compare with serving multiple shorter sentences for that same length of time? She suggested that bibliometric information be collected on surveys to get reliable information on health status of individuals instead of relying on self-reporting of health conditions and that latent-class analysis be used to tease out heterogeneous effects.

Ross Matsueda (University of Washington) asked Massoglia if there is any research on the heterogeneity of experiences within prisons. Massoglia said "no" and added that information on prison quality and prison experience is still sparse. Chris Wildeman (Cornell University) responded that departments of corrections have become more forthcoming over time in terms of sharing their data, and the hope is to explore questions on treatment heterogeneity.

## INCARCERATION AND HEALTH: EXPLORING NEW LINKAGES AND EMERGING VIEWPOINTS

This workshop session was aimed at considering the causal components of criminal justice involvement on health; identifying critical unanswered research questions and the data needed to explore them; and identifying optimal data collection strategies—that is, whether any of the proposed existing data collection instruments are able to meet the most critical uses given resource constraints.

Matsueda began the discussion by summarizing the viewpoints and information covered in earlier workshop presentations and raising other critical questions. For example: If one views incarceration as a treatment, what, then, is the counterfactual? How can one best deal with treatment heterogeneity? Matsueda asked the session participants to also address the issue of resource constraints on research.

Ingrid Binswanger (Kaiser Permanente Colorado and University of Colorado School of Medicine) focused on the principles that should guide decisions regarding collecting health information on the criminal justice involved population. From her perspective as a physician and health services researcher, Binswanger described the challenges in understanding the health outcomes of this population. She said that she considers the population to be "hidden" in public health surveillance and research: although the Bureau of Justice Statistics (BJS) collects statistics on the health of people in jails and prisons, they are largely excluded from major health surveys, despite being at high risk for many negative public health outcomes. Undercounting this population severely limits knowledge about their access to quality health care, as well as the ability to obtain accurate public health estimates.

Binswanger listed 10 important reasons that the bidirectional linkage between criminal justice and health needs to be more deeply explored:

- surveillance—improving measures of prevalence and incidence of epidemics, illness, and other emerging health issues;
- alignment—providing corresponding data on what is available on external populations in order to make appropriate comparisons and measure quality gaps;
- policy—addressing emerging policy questions, for example, gauging the impact of the Affordable Care Act on health care for criminal justice involved populations;
- equity—measuring variability, mortality, or other disparities by race, ethnicity, gender, geographic location;
- prevention—measuring the availability and delivery of preventive services (such as vaccinations);
- patient centeredness—assessing the quality of care to ensure that it responds to each individual patient's needs;
- transparency—providing accurate information about treatment, allowing public participation, promoting accountability;
- operations—using relevant data to guide criminal justice health care delivery, organization, and administration (such as human resources and contracting);
- performance and value—focusing on the value of care; and
- human rights and legality—assuring that patient rights are respected and constitutional or legal mandates are met.

Nagin highlighted the challenge of determining and measuring counterfactual outcomes when estimating the causal impact of incarceration on health. When one measures the effects of incarceration, to what condition is it being compared? What is the "alternative" treatment? In clinical trials, the control and treatment groups are well defined. When studying incarceration, Nagin encouraged the group to think about the best comparative treatment. The possibilities include individuals who were convicted but not incarcerated; individuals who were arrested but not convicted; and individuals who were not arrested. For more serious crimes, such as murder, incarceration post-conviction is imminent. However, for misdemeanors, summary offenses, and what Nagin refers to as "marginal felonies," the outcomes can vary greatly.

Nagin also picked up a point from Massoglia's presentation (see Chapter 4), noting that when comparing treatment during incarceration to treatment when not incarcerated it is important to consider both the health condition and the quality of treatment in both situations. There are some situations for which an individual may benefit from the access to health care that is provided in the correctional institution. These types of treatment heterogeneities merit serious exploration.

Evelyn Patterson (Vanderbilt University) talked about the collateral consequences of incarceration and how the effects of incarceration can reverberate throughout a person's life and the lives of his or her family and friends. She proposed a theoretical framework to understand the life experiences of people who have been incarcerated that highlights three major components: age, duration, and spells.

- Age: As people go through life they assume new social roles, some of which are cumulative—being old enough to work, to vote, becoming parents, etc. These roles are each affected when a person is incarcerated. Whether or not these roles are maintained during incarceration can provide insight into the effects of the situation on the individual and his or her family.
- Duration: Duration is also closely tied to social roles, in that the length of incarceration can be inversely proportionate to an individual's ability to preserve his or her social roles. How well can the incarcerated individual cope with the incarceration, or weather the storm, so to speak?
- Spell: A person's proximity to the incarceration experience affects how well he or she is able to understand it. Patterson pointed out that there are communities where the prevalence of incarceration is low and therefore general understanding of the experience and its effects are low.

Patterson noted that people's experiences and outlooks can be colored by their social, physical, and civic development and involvement. For example, although a person may be physically healthy, he or she may not have a fully formed civic or social life, which would hamper his or her chances of or expectations for upward mobility. Because this theory applies both to people who have and have not been incarcerated, it is possible that the quality of these theoretical lives can be similar across the two groups and not just within them.

Massoglia asked Nagin to elaborate on how predictive measures can help researchers hone in on the most important treatment variables for the criminal justice population. Nagin spoke about the benefits of longitudinal surveys like the National Longitudinal Study of Adolescent to Adult Health and how early waves of the study can inform researchers about the health statuses of candidates as they prepare for data collection in later waves. Binswanger commented that she still has difficulty processing incarceration as a treatment, since she sees it as a much-overused last resort for medical conditions (addiction, mental health) that could best be addressed in other settings. Nagin responded that, scientifically, the concept of treatment effect comes into play when one considers which treatments are necessary to make an inference that being sent to jail has a deleterious effect on an individual's health.

Josiah Rich (Brown University) picked up on Binswanger's mention of incarceration being improperly used as a treatment for addiction, mentioning that the lack of actual treatment for the problem often leads to a person's regression (and often overdose) when released. Rich suggested that Binswanger's 10 points be revisited with a focus not just on reducing incarceration, but also on doing so in a way that will be practical from a political and public acceptance standpoint, while also highlighting the beneficial effects and minimizing the detrimental effects of reduced incarceration.

John Laub (University of Maryland) asked Evelyn Patterson to offer an alternative strategy for collecting better measures of incarceration and health. Patterson said that she believes the key is to view incarceration as a social institution, exposure to which can affect the civic and social well-being of those currently and formerly incarcerated. She noted that health surveys do have questions that can measure social embeddedness or civic engagement. Including questions on incarceration in national health surveys can help researchers in parsing out the effects of incarceration, especially the compounding effects of incarceration spells.

Rashida Dorsey (U.S. Department of Health and Human Services [HHS]) thanked Patterson for her suggestion and reiterated one of the objectives of the workshop—to formulate criminal justice involvement questions that can be fielded in HHS population health surveys. The limitation of this approach is that these surveys are not given to currently incarcerated individuals, and the percentages of the survey samples that have had any criminal justice contact are very small. Patterson responded with the recommendation that questions be asked not just about a respondent's experience, but also about the household's exposure to the criminal justice system—i.e., proxy questions. She added that cognitive testing is required before fielding criminal justice questions in a health survey to understand respondent comprehension and recall.

Matsueda asked workshop participants who have experience conducting surveys and analyzing survey results to comment on what key data points are missing and what criminal justice experience questions should have priority for inclusion. Amanda Geller (New York University) pointed out the utility of linking survey data to administrative records. Speaking from her experience with the Fragile Families Study (see Chapter 7), getting consent from survey respondents to link their data to administrative records will complement the information from surveys. Administrative records are better at documenting duration and spells of incarceration than a survey, which relies on respondents' memory.

Christopher Uggen (University of Minnesota) referred to Wendy Manning's presentation (see Chapter 2) that highlighted the importance of disaggregating jail and prison experiences. Jail time may not substantially alter a life trajectory, while incarceration in prison can have a myriad of adverse effects, including loss of employment and increased risk of suicide and, for juveniles, increased risk of victimization. He supported Patterson's suggestion that duration is a key factor, and questions should be crafted in a manner that will aid people's recollection. Manning (Bowling Green State University) countered that while the two incarceration types differ vastly in design, prison overcrowding has resulted in longer jail stays for people who are convicted. And while jails might be less restrictive than prisons, they lack the structure and programs that are offered to prison inmates.

Following on this point, Laub suggested that researchers consider incarceration in prison and its recency and duration first, followed by filter questions on jails and probation. He mentioned the need to better understand prevalence and provide estimates of effects of any of these experiences on health. Cantor suggested that respondents may have an easier time recalling the duration of the stay than they will in differentiating their stays in a jail or a prison. Binswanger emphasized the importance of understanding the effects of the transition to different levels of criminal justice involvement when considering health outcomes.

The workshop participants discussed opportunities for combining resources and data linkages across surveys. Wang suggested that, as a start, surveys could ask about arrest history and compare the response rate on that question to the incarceration rate. She also reminded workshop participants that although surveys like the National Health and Nutrition Examination Survey (NHANES) do not specifically ask criminal justice questions, they do ask items that provide information on other social determinants of health, such as enrollment in the food stamp program. By elaborating on current questions (e.g., asking if a family has ever been banned from food stamps),[2] one can get information on the collateral consequences of criminal justice involvement.

Ann Carson (U.S. Department of Justice) described her agency's efforts to collect health data on incarcerated populations through collaboration: with the National Survey of Drug Use and Health, by including modules on risk, family, and social engagement for those who answer yes to probation or parole question; with the National Center for Health Statistics, to change the National Death Certificate Database to report if a death occurred in a

[2]According to the Pew Charitable Trust Legal Action Center, 6 states impose a full ban on food stamps for individuals convicted of a drug felony, and 24 other states impose a partial ban. (The District of Columbia and Puerto Rico were not included in this measure.)

correctional facility. She added that BJS has redesigned the inmate surveys to include modules on mental health.

Nagin asked survey practitioners if physiological measures can be collected from prison inmates. Kathryn Porter (National Center for Health Statistics) and Carson described the logistical challenges in collecting biological specimens in criminal justice institutions: NHANES uses mobile clinics; BJS relies on health care facilities in jails and prisons.

Jordyn White (Committee on National Statistics) reminded the workshop participants that space on existing national health surveys is limited, and as such the number of criminal justice questions that can be realistically included may well be less than what Wildeman proposed. She also encouraged participants to think about question priorities: a question asking about currently incarcerated family members may be more pertinent than one for respondents and other current household members who have had criminal justice experience, as an absent family member has a more pervasive effect on the family.

Massoglia pointed out that the desired answers will determine the proposed questions. Given resource limitations, it is important to understand what effects are of most interest to the research and policy communities. The questions need to be phrased according to research and policy priorities. Dorsey and Alexis Bakos responded that their agencies aim for better understanding of individual, family, and community effects, in that order of priority. With the awareness that formulation of question(s) depends on the constitution of the survey and respondent burden, the agencies are open to the possibility of rotating modules and linking across surveys.

# 5

# Asking Criminal Justice Involvement Questions

Household surveys can serve as a vehicle for collecting information on the experience of incarcerated people and measuring the effects of criminal justice contact on individuals, families, and communities. Because criminal justice involvement questions are sensitive, they can have an impact on survey response rates. It is a challenge to craft sensitive questions that will obtain the desired information without offending respondents. The workshop session covered in this chapter responds to the objective of developing ideas for potential survey questions that can be included in the household surveys of the U.S. Department of Health and Human Services. The goal was to provide examples of how sensitive questions are asked in surveys and discuss the potential impact they have on response rates. Both presenters in the session stressed the importance of cognitive testing and field testing sensitive questions before adding them onto a survey. Respondents with varied experiences approach the same question differently. Cognitive interviews can reveal those nuanced differences, which will improve context setting and framing of questions. Field tests provide an opportunity for interviewers to debrief respondents about content and ensure they understood what was being asked.

## ASKING SENSITIVE QUESTIONS

Ting Yan (Westat) opened her presentation with an example of different ways in which very sensitive questions can be presented to respondents. She referred to a classic paper by Allen H. Barton (1958) on how to ask embarrassing questions in non-embarrassing ways. For his study, Barton provided the following variations of the question, "Did you kill your wife?" as examples of strategies a survey might use to generate responses to difficult-to-answer questions:

- The casual approach: "Do you happen to have murdered your wife?"
- The numbered card approach: "Would you please read off the number on this card which corresponds to what became of your wife?"
- The "everybody" approach: "As you know, many people have been killing their wives these days. Do you happen to have killed yours?"
- The "other people" approach: "Do you know any people who have murdered their wives? How about yourself?"
- The sealed ballot approach: In an effort to preserve anonymity, the respondent is asked to place his or her response in a sealed ballot.

Yan said that questions are most often considered sensitive for one of three reasons: (1) they are deemed intrusive, asking people about details respondents feel are too personal to answer; (2) they cause concern for respondents about the threat of disclosure and repercussions if they respond truthfully; or (3) they raise social desirability concerns as respondents want their answers to reflect societal norms, which leads to overreporting of socially desirable answers and underreporting of socially undesirable answers.

Sensitive questions can be categorized into two types—those that are sensitive to ask and those that are sensitive to answer. Questions that are sensitive to ask are generally sensitive for all people and driven by content. Examples include questions on income, sexual partners, and Social Security numbers. Irrespective of wording and placement of such questions, they evoke the same level of sensitivity concerns among respondents. Questions that are sensitive to answer are perceived so by some respondents because of specific behaviors or attitudes. Examples include questions on use of marijuana and voting preferences.

Inclusion of sensitive questions in a survey can often result in a low participation rate, more missing data, and high measurement error. People, particularly those with undesirable behavior or attitudes, are less likely to participate in surveys with sensitive topics (Tourangeau et al., 2010). Even when people agree to participate in such a survey, they are more likely to not answer sensitive questions. For example, Yan and colleagues (2010) showed that income questions have 25 percent item nonresponse rates on surveys. Moreover, when responses are provided to sensitive questions, they frequently have measurement error from the usual sources (comprehension, retrieval, judgment), which is further compounded by self-editing done by respondents due to the nature of the question.

Yan discussed her research findings on items deemed sensitive by survey respondents. She and her coauthors surveyed University of Maryland alumni about various aspects of academic performance. Disaggregating survey responses by characteristics of people who participated in the survey, the researchers found that people who graduated with honors were more likely to participate in the survey than those who did not graduate. On questions of the students' status at graduation, the researchers found that survey respondents overreported graduating with honors and underreported dropping classes. However, the researchers found grade point average to be a more neutral item, having very minimum bias.

Yan refers to the process of providing an inaccurate response as "editing" and said that editing is a deliberate process even though sometimes it happens automatically or subconsciously. She described characteristics of respondents who are more likely to edit their responses: those with things to hide, those with a higher level of social security concerns, those perceiving survey items as posing a higher level of threats or losses are more likely to edit. For instance, literature on voting has shown that educated people are more likely to overreport having voted, as voting behavior is expected in their particular socioeconomic group (see, e.g., Preisendörfer and Wolter, 2014). In asking teenagers about smoking, Aquilino and colleagues (2000) found that they were more likely to respond affirmatively when interviewed in school settings than when interviewed in home settings, confirming that the presence of certain people encourages self-editing of answers. Also, Yan said, the more sensitive the question, the higher the likelihood of editing. Hindelang and McDermott (1981) found evidence of this phenomenon when measuring engagement in delinquent or criminal behavior.

Yan noted that a question on criminal justice involvement, irrespective of how it is defined, worded, and placed, is sensitive because of all three aspects noted above—intrusion, disclosure threat, and negative social stigma. In one study of criminal justice involvement, Preisendörfer and Wolter (2014) found some evidence of reporting bias in a mail questionnaire sent to people with some criminal history (they relied on criminal records as a sample frame). The authors found that women, older people, people with relatively higher education, people with higher social desirability concerns (measured through the survey), and people with more severe offenses were more likely to underreport their criminal histories. Underreporting was also found in people who responded late to the survey and those who were interviewed by less experienced interviewers. Overall, criminal justice involved populations are difficult to capture in household surveys as they are hard to find, and even if they are found, they are very likely not to report completely truthfully, Yan said.

Yan next turned to strategies to reduce underreporting of socially undesirable behaviors or attitudes: use of self-administered modes; question wording; question format; and indirect techniques.

**Use of Self-Administered Modes** Self-administered interviews reduce social desirability concerns and improve reporting. The National Survey of Family Growth (NSFG) and the National Survey of Drug Use and Health start with computer-assisted self-interviewing (CASI): at the beginning, interviewers administer the questionnaire to the respondents, but transition the respondent to audio computer-assisted self-interviewing (ACASI) for more sensitive items. In the NSFG, women respondents were more likely to report abortions on self-interviews than in personal interviews. Table 5-1 details the empirical research that has compared self-administered and interviewer-administered survey modes. The fourth column in the table is the ratio of the estimate of the variable (second column) between the two modes. A ratio of more than 1.0 refers to more reporting in the self-administered mode. There is strong evidence that switching to a self-administered mode leads to an increase in reporting of socially undesirable behavior, and the result holds across all modes.

**Question Wording** Questions on criminal justice contact can be framed in a manner that encourages people to admit that they have had such an experience. An example is to use a forgiving introduction. To see its effect on reporting of academic cheating, vandalism, shoplifting, littering, illegal drug use, and drinking and driving, Holtgraves and colleagues (1997) used the following forgiving introduction: "Almost everyone has probably committed vandalism at one time or another." The forgiving introduction improved reporting of academic cheating, illegal drug use, drinking and driving, vandalism, and littering; it did not change the reporting for shoplifting. Yan said that setting up a permissive context can make it easier for people to respond truthfully in some cases; therefore, careful pretesting is required to frame questions appropriately.

Three other strategies with question wording can be useful. One is to include using familiar words instead of abstract terminology: "arrested" rather than "criminal justice involvement." Another is to pose the question presupposing the behavior: "In the past 10 years, how many times did you witness a crime?" rather than "Did you witness a crime in the past 10 years?" The third is asking questions about past behavior, as the perception of past behavioral attitudes can be less sensitive than that of a current behavior: ask "have you ever" before asking "are you currently."

**Question Format** Response format can aid respondents' retrieval of memory, Yan said. For example, when respondents are asked about the

**TABLE 5-1** Comparison of Self-Administration and Interviewer Administration on Estimates of Variables of Interest in Criminal Justice

| Study Source | Topic | Modes Compared | Ratio of Estimates |
|---|---|---|---|
| Preisendörfer and Wolter (2014) | Ever convicted | Mail vs. face to face | 1.16 |
| Lind et al. (2013) | Number of sexual partners | ACASI vs. face to face | 1.93 |
| Villarroel et al. (2006) | Same-gender sexual experience | IVR vs. CATI | 1.56 |
| Schober et al. (2015) | Ever smoked 100 cigarettes | Texting vs. CATI | 1.08 |
| Corkrey and Parkinson (2002) | Ever use marijuana | IVR vs. CATI | 1.26 |
| Tourangeau and Smith (1996) | Ever use cocaine | ACASI vs. face to face | 1.81 |
| Tourangeau and Smith (1996) | Ever use marijuana | ACASI vs. face to face | 1.48 |

NOTES: ACASI, audio computer-assisted self-interview; CATI, computer-assisted telephone interview; IVR, interactive voice response.
SOURCE: Yan (2016).

number of times or frequency of occurrence, use of a specific range or high frequency list (0, 1-4, 5-9, 10-59, 60-99, 100 or more) generates more responses than an open-ended format. Such a list makes the respondents speculate about the mean answer and rank order their answers relative to the mean, which improves the chances of getting a truthful answer as respondents believe they are doing better than average.

**Indirect Techniques** Indirect techniques that are helpful in understanding population sizes include randomized response techniques, crosswise models, and item count techniques.

## CAPTURING INFORMATION IN A FEW SIMPLE QUESTIONS

David Cantor (Westat) began his presentation by responding to Yan's point about error. He said that although there is surely error associated with asking the types of questions being discussed, research shows that the error rate for them is not significantly higher than for other questions. He noted that response error is most commonly due to a lack of question comprehension or a problem with recall and that it is possible to significantly reduce total error by carefully designing questions to minimize these common issues.

Cantor offered the example of the San Jose recall study, in which respondents were asked to report instances of victimization—namely, rape, assault, and robbery (Turner, 1981). These self-reports were then matched to police records. Though it is arguably the most sensitive question, the percentage of rapes that were accurately reported (67) fell between the percentages of assaults (48) and robberies (76). Cantor noted that the error rate was likely higher with the assault question because the respondents' comprehension of the term "assault" can vary greatly. He also noted examples from his 1996 survey of criminal offenders of how incident recall for topics like drug use can diminish greatly over time. Referring to question design advice from Sudman and Bradburn (1982), he offered guidelines on framing questions on sensitive topics, including making questions as specific as possible, using words that virtually all respondents will understand, and lengthening the questions by adding memory cues and examples to improve recall. He added that the guidelines can be compromised for particular applications, but specificity is key for ensuring comprehension. For example, when collecting data on sexual assault, it is better to ask whether respondents have had some sort of sexual contact against their will than to simply ask if they have been victims of assault. Breaking down complex items can help respondents better understand the question and attend to the specific terms. Cantor added that the work of Biderman and colleagues (1985) and Tourangeau and colleagues (2015) has shown that use of examples makes the memory search more specific to the behavior or situation that the researcher is targeting. He noted, however, that although such strategies can prove helpful, the tradeoffs are the additional space required to include examples and the concern that too much specificity can ultimately confuse respondents.

Cantor discussed whether the criminal justice contact questions on various national health surveys followed the guidelines. The Survey of Criminal Justice Experience (SJCE), the NSFG, and the 1997 National Longitudinal Survey of Youth (NLSY) all included a question about supervision that listed the specific levels of interest; for example SJCE asked, "Since age 18 have you ever been under any form of criminal justice supervision, including on probation, in jail, or in prison?" The National Longitudinal Study of Adolescent to Adult Health (Add Health), the National Survey of Drug Use and Health (NSDUH), and the NLSY all contain questions about criminal justice contact, but the questions vary greatly.

- Add Health asks respondents if they have ever been arrested or taken into custody by police. Asking about contact in this manner mitigates the ambiguity often seen when asking respondents what actually constitutes a formal arrest—being picked up by police, being interviewed, being formally booked, etc.
- The NSDUH includes a definition in the question to aid recall and accuracy: "Not counting minor traffic violations, have you ever been arrested and booked for breaking the law? Being "booked" means that you were taken into custody and processed by the police or by someone connected with the courts, even if you were released."
- The NLSY uses an exclusion phrase in the question in an attempt to weed out certain types of offenses: "Have you ever been arrested by the police or taken into custody for an illegal or delinquent offense (do not include arrests for minor traffic violations)?"

Drawing on earlier discussions during the workshop, Cantor outlined the essential informational needs of the research and policy communities and questions that need to be addressed before framing survey questions. He also provided examples of questions to obtain information about criminal justice contact (see Table 5-2).

In conclusion, Cantor said the most important principle is to first decide on the desired content. What is your variable of interest? What are you trying to estimate? Then be sure you are providing the right context of reference for survey respondents.

**TABLE 5-2** Possible Survey Questions on Criminal Justice Contact

| Information Sought | Possible Survey Questions |
| --- | --- |
| Type of contact: police contact, conviction, or supervision | Since you were 18, have you been arrested for breaking the law? Please don't count minor traffic violations.<br><br>Since you were 18, have you been convicted of a crime? Please don't count minor traffic violations.<br><br>Since you were 18, have you been locked up for at least 30 days to punish you for breaking the law? |
| Timing: recent versus lifetime | Specific calendar periods are often difficult for respondents to remember, as people tend to store life events in their memory by their significance.<br><br>Asking about experiences in a specific time frame—the past one or two years, for example—can cause respondents to bring their past experiences forward into the reference period in a phenomenon called telescoping. This can bias the estimates on prevalence. One can instead consider anchoring to specific events such as "most recent time it happened" or "age first occurred." |
| Number of times versus time period | Choose between asking people how many times an event happened and how frequently the event happened.<br><br>Another level of decision making happens when choosing between the question being open-ended or providing a response scale. |
| Length of time: total, by spell, or longest | Asking respondents about how long an incarceration term or spell occurred requires them to do calculation and estimation. As longest spells are generally easier to remember, the question can ask specifically about that. |
| Type of crime: specificity | Since you were 18, have you been arrested for any of the following (y/n to each):<br>• stealing something<br>• breaking into a home or a building<br>• assaulting someone with a weapon<br>• injuring someone with a weapon or your hands<br>• injuring someone with a motor vehicle<br>• drunk driving<br>• tricking a person into giving you money<br>• using someone else's identity to obtain money<br>• for anything else? |

SOURCE: Cantor (2016).

# 6

# Measuring Criminal Justice Contact and Incarceration Experience in Health Surveys

Michael Schober (New School for Social Research) began the workshop session by pointing out that each of the five health surveys that will be discussed has its own core functions and constituencies whose information needs it specifically addresses. There is wide variety in the sampling processes, sample sizes, administration modes, and areas of measurement across the surveys. He acknowledged that inclusion of criminal justice involvement measures on these surveys can have implications for instrument length, respondent burden, and, ultimately, response rates, which are already a concern for some surveys. Schober asked the workshop participants to also bear in mind that because of the large variation among the surveys, a singular set of proposed changes may not be feasible.

## THE NATIONAL HEALTH INTERVIEW SURVEY

Marcie Cynamon (National Center for Health Statistics [NCHS]) provided an in-depth overview of the National Health Interview Survey (NHIS).

NHIS is one of the largest and oldest health surveys overseen by the U.S. Department of Health and Human Services.[1] The survey is aimed at collecting and analyzing data on a broad range of health topics to monitor the health of the non-institutionalized civilian population. The NHIS is an address-based multistage probability sample that collects data from 75,000 to 100,000 individuals in 35,000 to 40,000 households. There are

[1]The U.S. Census Bureau serves as the data collection agent for this survey.

at least two primary sampling units in each state, which enables production of state-level estimates on a wide range of variables. NHIS is primarily administered in person in respondents' homes, with roughly 22 percent of the interviews completed by telephone.

Since the survey began in 1957, Cynamon noted, NCHS has produced public use microdata files and analytic reports. More detailed information that cannot be released on public use files due to disclosure risk may be available via the NCHS Research Data Center. The NHIS data are linked to other data sources, such as the National Death Index, and to data files from other agencies, such as the Centers for Medicare and Medicaid Services and the U.S. Department of Housing and Urban Development.

Cynamon explained that since 1997 the survey has included three modules, each directed toward different component in a household: the adult(s), the child(ren), and the entire family. The first 20 to 30 minutes are a family-level interview in which information about the entire household is collected and a knowledgeable respondent answers some health questions about everyone in the household. The next two segments are directed to one adult and one child (if there are children in the household), who are randomly selected, for a more detailed interview, each approximately 20 minutes long. In addition, the questionnaires have supplements that can pertain to the sample adult or, in some cases, to the sample child, which add another 20 to 30 minutes to a respondent's total interview time.

Currently, Cynamon explained, the survey is being redesigned to improve measurement of covered health topics, reduce respondent burden, harmonize overlapping content with other federal health surveys, establish a long-term structure of ongoing and periodic topics, and incorporate advances in survey methodology and measurement. As part of that redesign, NCHS plans to incorporate biomeasures through a passive monitoring approach. The redesigned questionnaire will be finalized by early 2017 and fielded in 2018. Outreach activities to survey users and professional associations have generated many suggestions on rearrangement of the survey, which the agency has accommodated to the best of its capacity.

Cynamon described how the trends in questionnaire length and response rates in the survey's three cores are moving in opposite directions. Declining response rates affect the reliability and validity of NHIS estimates. Instead of having an entire family module, the proposed new structure of the survey will have a brief set of questions about the family, framed in a manner to collect basic demographics about a household that will aid in selection of a sample adult and sample child. The focus of the interview will now be on the sample adult and sample child, reflecting that 95 percent of NHIS publications show data from these individual levels. The new adult core module contains information about family income and program participation. The content of the child core is still under development. However, the redesigned survey will still contain annually rotating core modules about additional household content, such as tobacco use and specific health conditions, and certain federal agencies will also have the opportunity to sponsor 5-minute-long supplemental modules.

Cynamon described the decision criteria of inclusion of questions or topics in the core module:

- leading causes of morbidity and mortality;
- health insurance coverage;
- health care access and utilization;
- targets of major federal health promotion initiatives;
- behavioral risk or protective factors for the above relevant content areas;
- other factors that identify populations most at risk for or protected from the above relevant content areas (e.g., family income); and
- intermediate health outcomes for the above relevant content areas.

## THE BEHAVIORAL RISK FACTORS SURVEILLANCE SYSTEM

Wayne Giles (Centers for Disease Control and Prevention [CDC]) gave an overview of the Behavioral Risk Factors Surveillance System (BRFSS), which is a system of state-based health surveys. The BRFSS has been in existence since 1984; it arose from a need for public health data at state and county levels. He noted that since the implementation of public health programs and health promotion often takes place at the state and county levels, it was recognized that there is a need for health surveillance data at those levels. The BRFSS is an outcome of a joint partnership between CDC and states, with the content decided jointly as well.

The state administers telephone surveys by both landlines and cell phones. More than 450,000 interviews are completed each year. On average, 35 percent of the interviews take place on cell phones, but in some states the rate is more than 60 percent. Survey respondents are asked about health conditions, health behaviors, and health care utilization. The BRFSS response rates for combined landline and cell phone interviews range between 42 and 80 percent (the median is at 67 percent). Giles noted that it is a challenge to get people to answer the phone, but once they answer, BRFSS interviewers are generally successful in conducting an interview. He also noted that interviewers have held people's interest on cell phones longer than on landlines.

Giles described the four components of the BRFSS. The core survey includes annual and rotating core sections and is uniformly fielded in all states and territories. Questions are asked every year on the core survey topics: HIV/AIDS, diabetes, asthma, cardiovascular disease, alcohol consumption, health status, health care access, healthy days, emotional support and life satisfaction, disability, tobacco use, and sleep. There are also rotating core topics. In even-numbered years the topics include breast and cervical cancer screening, prostate screening, colorectal cancer screening, oral health, falls, seatbelt use, and drinking and driving. In odd-numbered years, the topics are fruit and vegetable consumption, hypertension awareness, cholesterol awareness, arthritis burden, and physical activity. Socioeconomic characteristics and demographic information are also collected.

The BFRSS also fields supplemental modules that are generally proposed by CDC programs and other agencies, including the Substance Abuse and Mental Health Services Administration and the U.S. Department of Veterans Affairs. States also include specific questions to meet their individual needs and issues. Giles said that the BRFSS also allows for inclusion of special project questions, such as H1N1 flu. These add-on special project questions are included on an as-needed basis, generally in response to requests from various programs and agencies. Giles said that data from the BRFSS are used extensively by the federal and state health departments, policy makers, universities, researchers and research organizations, media, insurance companies, and the general public. One primary use is in the publication *America's Health Rankings*, which ranks states and counties on their core measures of health and health outcomes. Giles also mentioned a 500-city partnership project between the CDC Foundation and the Robert Wood Johnson Foundation to produce, analyze, and release census-tract-level health data compiled from multiple sources that will produce health outcomes data for neighborhoods and cities.

## THE NATIONAL SURVEY OF DRUG USE AND HEALTH

Arthur Hughes (Substance Abuse and Mental Health Services Administration [SAMHSA]) described the design, sampling frame, instruments, and response rates of the National Survey of Drug Use and Health (NSDUH) and discussed the potential of adding a criminal justice module to the survey. The NSDUH estimates are the primary source of statistical information on substance use and mental health issues in the U.S. civilian noninstitutionalized population. The survey is sponsored by SAMHSA and planned and managed by the SAMHSA Center for Behavioral Health Statistics and Quality, with data collected by RTI International. The NSDUH estimates are produced at national, state, and below-state levels, which helps determine substance use and mental health service needs and allocation of funding to meet those needs. The data are useful in gauging trends of substance use, substance use disorders, and mental health problems in the general population. The survey also collects information on such covariates as arrests, probation, and parole.

Hughes explained that the NSDUH is designed to be a multistage area probability sample of households. States are the first level of stratification, followed by stratification within each state, then census tracts, census block groups, and segments consisting of a collection of census blocks are selected within each state stratum. The 12 largest states have an annual target sample of over 1,500 each, while the 38 other smaller states plus the District of Columbia have a sample size of 960 each, which brings the total sample size to approximately 67,500 people. There are several different stages in the sample selection process. Geographic selection of segments is conducted in stage three, and these areas are canvassed to generate a list of dwellings (addresses/households). Survey statisticians select the sample dwelling units from the canvassed list. Once a dwelling is selected, a professional RTI interviewer conducts a screening interview, and then zero, one, or two people aged 12 and above are selected from the household for an in-person interview. The survey oversamples people aged 12-25. The response rate for the 2014 NSDUH screener was 82 percent, and the response rate for the in-person interview was 71 percent. The NSDUH relies on a computer-assisted personal interview (CAPI) for collecting demographic variables and on an audio computer-assisted self-interview (ACASI) for collecting data on recency, frequency, and initiation of substance use.

Hughes told the workshop participants that questions about criminal justice involvement were added in 1991 to address the paucity of data on the relationship between drug use and criminal behavior among the general population. These questions are part of the special topics module, which is administered to respondents aged 12 and older. This NSDUH module asks whether the respondent has ever been arrested and booked for breaking a law. If the answer is yes, the respondent is then asked how many times he or she has been arrested, and the respondent is also asked questions about the type of offense. The reference period for all of the criminal justice questions is the past 12 months.

The NSDUH module covers 17 offenses:

- motor vehicle theft;
- larceny or theft;
- burglary or breaking and entering;
- aggravated assault;
- other assault, such as simple assault or battery;
- robbery;
- forcible rape;
- murder, homicide, or non-negligent manslaughter;
- arson;
- driving under the influence of alcohol or drugs;
- drunkenness or other liquor law violations;

- possession of tobacco (only for respondents aged 12-17);
- possession, manufacture, or sale of drugs;
- prostitution or commercialized sex;
- any other sexual offense, not including rape or prostitution;
- fraud, possessing stolen goods, or vandalism; and
- some other offense.

The survey also includes questions on probation, parole, supervised release, and other conditional release with a recency period of 12 months.

Hughes reported that from 2008 to 2014 the average weighted prevalence estimate of respondents who had ever been arrested was 16.8 percent. For probation the prevalence estimate was 2.2 percent, and for parole/supervised release it was 0.7 percent. The total sample size of ever arrested over the same 7-year period was 67,000. The numbers for on probation and on parole/supervised release were 15,000 and 4,400, respectively. Less than 1 percent of data are missing for these variables. The average questionnaire time of NSDUH 2015 was 58.7 minutes for all respondents. However, the response time was 63.9 minutes for respondents who answered yes to the probation question, and was longest for respondents who answered yes to the parole/supervised release question: 66.4 minutes.

Hughes discussed the need to be mindful of question placement when considering adding additional criminal justice questions to a survey like the NSDUH. Currently, the NSDUH does not collect any data related to criminal justice other than prevalence and timing. Adding questions at the end of the special topic module would be difficult given the structure of the interview, which moves from CAPI to ACASI then back to CAPI and identification of proxy respondents for selected questions. Adding questions elsewhere may produce context effect to other questions, including sensitive back-end CAPI questions.

Hughes also noted that placement of new questions on the interview would also undoubtedly produce context effects. And since the population of interest among the NSDUH respondents is small, field tests may not accurately detect any such effects and their consequences. Moreover, since the survey asks sensitive questions, additional questions may raise concerns about approval from an institutional review board (IRB).[2] He suggested that the optimal approach toward implementing a new module on criminal justice would be to incorporate it into the next NSDUH redesign given the assumption there will be an understanding that the overall change will create a new baseline.

[2]An IRB has to approve any sensitive survey questions that involve human subjects (respondents).

## THE NATIONAL SURVEY OF FAMILY GROWTH

Anjani Chandra (National Center for Health Statistics) provided a brief description of the National Survey of Family Growth (NSFG) and discussed the incarceration questions it includes for male respondents.

The NSFG came under the federal statistical system due to a portion of the Public Health Service Act, which stipulated that the agency should collect statistics on family formation, growth, and dissolution. The survey is primarily funded by NCHS, the National Institute of Child Health and Human Development, and the Office of Population Affairs, all in the U.S. Department of Health and Human Services. The primary purpose of the survey, which has been conducted since 1973, is to explain variation in birth rates. Thus the survey context is organized around proximate determinants of fertility, looking at factors that directly affect the occurrence of sexual activity, the occurrence of pregnancy from sexual activity, and the occurrence of a live birth from pregnancy. The survey includes items on sexual activity, contraception, infertility, pregnancy loss, and other topics that would affect occurrence of a birth. The NSFG also collects information on covariates such as demographic characteristics and socioeconomic factors.

Chandra explained that the first two rounds of the survey were targeted at ever-married women; then, in 1982, the survey expanded to encompass all women aged 15-44. In 2002, the survey scope was broadened to include men of the same age range, and the content was also expanded to include risk factors for HIV, sexually transmitted diseases, and nonmarital fertility. In addition to complete pregnancy histories and marriage and divorce information, the survey collects data on the roles of fathers in raising children and attitudes about sex, marriage, and parenthood with an aim to understand and describe family formation, growth, and dissolution. In 2015 the age range was expanded to 15-49.

Chandra pointed out that in its current form the NSFG is not a survey of couples: the sample of men is independent from the sample of women. In 2006, the NSFG transitioned from a periodically conducted survey to a continuous fieldwork design, which increased the number of interviews by 80 percent for about one-third of the cost. The continuous design has also enabled release of public-use data files every 2 years. The 2013-2015 public-use file was released in fall 2016.

Chandra described the steps involved in selecting an NSFG sample. First, primary sampling units (PSUs) are selected, consisting of Metropolitan Statistical Areas, counties or groups of counties, and then neighborhoods and secondary sampling units are selected from within those PSUs. A household screener is conducted for selected households to locate eligible men and women ages 15-49 for interview (until September 2015 the range was 15-44). The interviewers are women and are trained for a week exclusively on the NSFG (including refusal training). The average interview length is 80 minutes for females and 60 minutes for males; the majority is interviewer administered on a laptop, but 15 to 20 minutes of each interview is self-administered using audio content (ACASI). Chandra mentioned that the latter portion contains their most sensitive content, including their questions on incarceration. Parents of NSFG respondents aged 15-17 must provide assent themselves and sign parental consent. Chandra presented a summary of the questions on criminal justice involvement in the NSFG: they are administered only to men, in ACASI mode. The questions are framed in a manner so as not to be perceived as self-incriminating. The question on "number of times" is aimed at distinguishing serious offenders from those who have committed less serious crimes:

*In the last 12 months, have you spent any time in a jail, prison or a juvenile detention facility? (yes/no)*
If no to the above question:
*Have you ever spent time in a jail, prison or juvenile detention center? (yes/no)*
If yes to either jail question on prior slide:
*Have you been in jail, prison, or a juvenile detention facility only one time or more than one time? (only 1 time; more than 1 time)*
If only 1 time:
*How long were you in jail, prison, or juvenile detention?*
If more than 1 time:
*The last time you were in jail, prison, or juvenile detention, how long were you in?*
Response options: One month or less; more than one month but less than one year; One year; more than one year

## THE NATIONAL HEALTH AND NUTRITION EXAMINATION SURVEY

Kathryn Porter (National Center for Health Statistics) described the National Health and Nutrition Examination Survey (NHANES), which is conducted by NCHS. NHANES is a nationally representative annual survey of a sample of about 5,000 people focused on assessing the health and nutritional status of adults and children in the United States. Unlike the four surveys described above, the NHANES combines interviews and physical examinations. The data are used to produce U.S. population-based estimates of health conditions, environmental exposures, nutritional status, and diet behaviors, as well as to assess the awareness, treatment, and control of selected diseases. In the 2015 round, the survey's response rate was around 63 percent, which Porter noted was a disappointing decline from a decade ago in which response rates were in the high 70s.

Like the previously mentioned health surveys, Porter explained, the NHANES is administered to a sample of the civilian, non-institutionalized U.S. population. The sample is selected in four stages: counties are selected as PSUs: segments or groups of census blocks are selected within each PSU; households are selected within each segment; and participants are selected from households through screening and enumeration. Informed consent is obtained prior to both the interview and the physical examination. The consent includes language regarding the linking of NHANES data with other survey records and administrative data. Respondents are therefore agreeing not just to provide their personal data, but also to allow NCHS to link their personal data to other sources.

The home interview includes demographic, socioeconomic, dietary, and health-related questions. The examination component consists of medical, dental, and physiological measurements, as well as laboratory tests administered by trained medical personnel. The collected biological specimens are either banked or sent to laboratories for processing. Physical examination takes place in mobile examination centers that are equipped with medical devices to assess cardiovascular health, conduct exams for sensory impairments, and collect anthropometric information. Sensitive questions, such as those on mental health, reproductive health, drug use, and alcohol use, are asked of participants in the mobile examination center.

Porter informed the workshop participants about the procedure to add new content to the NHANES. The program openly invites new or revised questionnaire material, laboratory assessments, and examination contents from current and potential future collaborators from federal and state governments, the private sector, and universities. There is a 2-year timeline for incorporating suggestions, which starts with a letter of intent, followed by clearances from the U.S. Office of Management and Budget and the relevant IRB, cognitive testing of proposed questions or assessments, and training of interviewers, ending with fielding of new content.

## DISCUSSION

Schober thanked the session speakers and opened the floor to discussion by requesting, first, that the presenters comment on the discussions and presentations earlier in the workshop.

Hughes talked about the tradeoffs involved in the redesign of a survey and how context effects can affect the estimates and distribution of variables. Context effects can create disruption in the general picture of how things are moving in a particular direction. In case of the NSDUH, he suggested expansion of the current criminal involvement module rather than placing new questions in a separate section of the survey—primarily because the criminal justice module is already designed to be administered by ACASI, and the respondent is already privately answering questions.

Hughes then turned to some examples of situations in which additional questions on a survey generated inconsistent responses that rendered the data unusable. One such example was the result of a change in definition of binge drinking for women by the National Institute on Alcohol Abuse and Alcoholism from five or more drinks to four or more drinks. Instead of redesigning the alcohol module, a new question that took into account the changed definition of binge use was introduced in another module so as to not disturb trends in estimates: as a result, however, it generated inconsistent responses. The data from the new question became usable only after the questions in the alcohol module were redesigned to reflect the changed definition.

Jordyn White (Committee on National Statistics) asked when criminal justice questions were introduced to the NSDUH. Hughes answered that they were introduced in 1991 and were very basic, asking whether the respondent has ever been arrested, been on parole, or on probation. In the mid-1990s, NSDUH partnered with the Bureau of Justice Statistics, which led to the inclusion of questions on specific offenses (the list is drawn from Federal Bureau of Investigation's Uniformed Crime Report list of offenses) to generate estimates of their prevalence. The estimation was a challenge given the sample size of the survey.

Cynamon followed up on Hughes's answer by reminding workshop participants of the drop in response rates in household surveys. She said she is especially concerned about the fact that the demographic group with the largest nonresponse rate is young adult minority males, which affects the total number of observations from that group and in turn leads to issues of undercoverage and biased estimates.

Chandra reiterated the need to balance challenges when adding new questions to a survey. First, there are concerns that they can cause disruption in time series of variables that have been in core modules since the beginning of the survey. But by placing additional questions in a separate module to protect the content of the core module, the additional questions do not derive the benefit of the context set up in the core module. She noted that NSFG's timeframe for making questionnaire changes is very similar to that of the NHANES: the survey is currently considering changes that would be implemented in September 2017.

Giles relayed concerns from BRFSS coordinators regarding the length of the current questionnaire, which would therefore make adding criminal justice content in BRFSS's core module all the more difficult: it would either increase respondent burden and length or require the survey to remove a current item. He agreed with the other presenters' concerns about the effects of context and validity. The BRFSS's rotating modules' placement on the questionnaire does not vary from one year to the next—a decision that helps to protect the integrity of the core module. He added that the program will begin soliciting suggestions for content modifications and additions for the 2018 BRFSS in fall 2016, which would be the earliest opportunity to add criminal justice questions in the BRFSS.

## ADDING QUESTIONS TO HEALTH SURVEYS: ADDITIONAL CONSIDERATIONS AND CONCERNS

Josiah Rich (Brown University Medical School) thanked the presenters for describing five important national health surveys. He challenged the presenters to think about three possible approaches:

1. linking their respective survey data to the National Correctional Reporting Program's national database of prisoners and the Federal Bureau of Prisons database, which can generate data on prevalence and help validate questions regarding incarceration and health;
2. expanding the coverage of their surveys to include representative subsamples of correctional populations, which would aid in a comparison of health outcomes of the correctional populations relative to general populations; and
3. including content on opiate use disorder and overdose.

Responding to Rich's third point, Porter described results of analysis done on prescription drug data for 10 years from the NHANES: there has been an increase in prescription opioid use, from 4 percent to 6 percent. In addition, the proportion of people taking opioids stronger than morphine has greatly increased. She noted that conducting physical examinations for people in correctional institutions would require changes in the authorizing legislation for the NHANES. An alternative would be to share NHANES methods and protocols with a private-sector contractor that could go into jails and prisons and collect biological specimens. She gave an example of the collaboration with the New York City (NYC) Department of Health and Mental Hygiene to design and implement a community HANES. The NYC HANES measured key health indicators in a randomly selected sample of adult residents through a detailed health interview and brief physical examination.

However, Porter said, she has reservations about linking NHANES survey data to administrative data on correctional and prison population because the annual NHANES sample size is only 5,000, and a very small percentage of respondents have any contact with the criminal justice system. She opined that aggregating 10 years of survey data may produce a sufficient sample size that can be linked with databases—one of the options Rich suggested—but for such an exercise to bear fruit it would require formulation of research questions and investigation of statistical power of the survey to provide answers to those questions.

Regarding the NSDUH, Hughes said that when the survey expanded the prescription drug module significantly by adding more drugs, the usage data broke prior trends. While there had been a general decline in prescription opioid use, the frequency of usage among heavy opioid users has been increasing. In responding to Porter's point that survey mandates do not cover correctional populations, Hughes said he was not sure if there is added value in linking NSDUH data to databases on correctional and prison populations.

# 7

# Alternative Approaches to Criminal Justice Measurement

John Laub (University of Maryland) began the workshop session on looking at alternative approaches by summarizing the strategies proposed earlier in the workshop for collecting information on the health effects of criminal justice involvement. Those strategies included linking criminal justice involvement to ongoing general social surveys, gaining more knowledge of the needs of the incarcerated population through criminal justice and inmate surveys, and using administrative records to make linkages to past incarceration experience. He then turned to the designated speakers to discuss their experiences with the Fragile Families Study, community-based surveys, the National Longitudinal Survey of Youth, the Panel Survey of Income Dynamics, and the National Crime Victimization Survey.

## MEASURING INCARCERATION IN THE FRAGILE FAMILIES STUDY

Amanda Geller (New York University) described measures of incarceration and criminal justice involvement collected in the Fragile Families and Child Well-Being studies and what has been learned from a pilot linkage project with administrative data provided by the New York State Division of Criminal Justice Services.[1]

[1]Administrative data were provided by the New York State Division of Criminal Justice Services (DCJS) in the interest of information exchange. The opinions, findings, and conclusions are not necessarily those of DCJS. Neither New York State nor DCJS assumes liability for these findings or the use thereof.

Fragile Families is a birth cohort study of 4,898 families with children born in 20 large cities (populations of 200,000+) between 1998 and 2000. The study was originally designed to primarily address four questions of great interest to researchers and policy makers: (1) What are the conditions and capabilities of unmarried parents, especially fathers? (2) What is the nature of the relationships between unmarried parents? (3) How do children born into these families fare? and (4) How do policies and environmental conditions affect families and children? The study has since expanded to look at broader social disadvantages faced by children (e.g., paternal incarceration) and is currently in its sixth wave of data collection, surveying children 15 years of age. As nonmarital births are systematically oversampled in the study, about 75 percent of the children in the study were born to unmarried parents, which resulted in a sample dominated by children belonging to minority racial and ethnic groups of lower socioeconomic status and high rates of paternal incarceration (about 50 percent). The core study consists of interviews with both mothers and fathers at birth and again when children are 1, 3, 5, and 9 years old. At the time of writing, the year 15 survey focuses on the children and their primary caregivers, and does not interview fathers.

As a population-based survey rather than a survey of inmates, Fragile Families collects information on incarcerated and formerly incarcerated parents, as well as on parents without histories of incarceration experience. The systematic oversample of unmarried parents in the Fragile Families sampling plan produces a sample that is highly socioeconomically disadvantaged, with high rates of incarceration experience, particularly among the fathers. This makes the study of parental incarceration more feasible than many other samples. Fragile Families also provides ways to compare fathers who have an incarceration history to those who do not, but who nonetheless experience very similar socioeconomic disadvantages. Incarceration can be reported in self-reports or in partner reports, providing some information about incarcerated parents lost from the survey due to attrition. Incarcerated fathers are often non-custodial parents and not the primary caregivers, which makes it hard to identify father–child relationships in other administrative datasets (such as incarceration surveys). Fragile Families provides an opportunity to link currently and formerly incarcerated fathers to their children.

Geller noted that both parents are asked questions about their own and their partner's involvement with the criminal justice system. Generally, fathers are asked more detailed questions than mothers. They are first asked if they have been charged with a crime in a particular time period; if so, they are asked if they were convicted of a crime; and if so, they are asked if they have spent time in prison or jail. Mothers, on the other hand, are simply asked if their partner has ever been incarcerated. Mothers' questions are therefore more inclusive than fathers': If a father was charged with a crime, held in central booking or jail for a period of time, but charges were ultimately dismissed, mothers would have an opportunity to report the incident, but fathers would not: they would report the charge, but with no conviction, they would not be asked about incarceration. In addition there are questions that gauge criminal justice involvement indirectly, such as asking why the mother and father are separated, or why one parent has been unable to search for employment, providing incarceration as an answer choice. Geller said that her analysis shows that the majority of incarceration experiences are identified through partner reports rather than self-reports, since response rates are higher for mothers than fathers (Geller et al., 2016). She noted that most incarceration analyses based on Fragile Families assume that a self-report or partner-report of an incarceration incident is correct, since research shows that potentially stigmatizing behavior is unlikely to be overreported (see Chapter 5 on how social desirability influences answers to sensitive questions).

Geller and colleagues (2016) explored the possibility of improving Fragile Families data quality by supplementing it with administrative records on criminal histories. Official criminal history information may identify and provide more detail on incarceration experiences that are unreported or underreported in survey data. Conversely, a data linkage exercise such as this can also reveal to what extent official records miss criminal history experiences that are reported in surveys. As part of the data-matching pilot study, information on 333 fathers from the Fragile Families New York City sample was submitted to New York State's Department of Criminal Justice Services after working to design a data collection protocol that adequately addresses human subject protection and privacy protection. Records were matched on the basis of name, Social Security number, and date of birth. Analysis of the data showed a 23 percent increase in fathers identified as incarcerated, and a 30 percent increase in fathers identified as criminal justice involved (Geller et al., 2016).

Geller stressed that findings from the data linkage exercise underscore the challenges in identifying criminal justice involvement in surveys and the extent of criminal histories that do not show up in administrative records. For example, criminal history records may be sealed under a variety of circumstances, making it difficult to collect full information on criminal histories. The exercise also makes clear that administrative records are not a panacea for underreporting in surveys because of their aforementioned shortcomings and access restrictions.

Geller provided directions for future research in the area of survey methodology and data linkage, drawing on experience from the data matching pilot:

1. Collecting administrative data prospectively alongside survey waves could make matching more manageable.

    - Respondents could provide consent for comprehensive administrative record searches during the survey itself, so that human subjects concerns are addressed prospectively.
    - Survey questions could be used to identify states to search records, increasing the probability that administrative records fully reflect respondents' criminal histories.
    - Collecting data at every wave can help ascertain timing.
    - This prospective collection still would not, however, include sealed records in states that cannot release them.

2. Details provided in administrative data raise questions about how respondents might interpret survey questions, which may be addressed through testing or qualitative research.

    - Do respondents count "convictions after plea bargains" as convictions?
    - How about adjournments in contemplation of dismissal and pleas to violations (e.g., disorderly conduct)?
    - Do respondents count pretrial detention as incarceration?

3. Future surveys might include cognitive tests in which respondents randomly receive alternate question wordings, to see if more explicit explanation leads to higher reporting rates.

## COMMUNITY-BASED SURVEYS

John Boyle (ICF International) drew on his experience conducting more than 3,000 surveys to discuss how household surveys can serve as a vehicle for collecting information on such topics as incarceration experience and criminal justice involvement.

Boyle suggested that a cross-sectional survey may not be the best way to measure the variables that are of interest: instead, workshop participants might consider an age cohort study in which individuals are followed over time. He cited the example of the Longitudinal Harlem Adolescent Health Study, a community-based sociomedical study focused on the health problems of teenagers. It was a community-based study fielded in the late 1960s in which participants aged 12-17 were selected from one in six households in the central Harlem health district. The study participants were followed for 25 years, with their information collected in five waves. A general household survey may not have sufficient sample size to answer questions on the health effects of incarceration, but a community-based study such as this one, with its focus on a high-risk community, will have enough study participants who have gone through an incarceration experience or had some other level of contact with the criminal justice system to serve this purpose.

The second advantage of an age cohort study, Boyle said, is twofold; it controls for age, and it also includes individuals living under the same conditions who did not have such an experience, thereby facilitating the definition and determination of counterfactual cases. Such an approach requires careful selection of households so that study participants remain engaged with the study for a long period of time. A community-based cohort study allows for logistical support to collect biological specimens because the location is fixed, and interviewers are going back to the same household and location in every wave. Although longitudinal surveys are excellent for studying health outcomes of criminal justice involvement, Boyle acknowledged that they are costly and require enormous amounts of time.

Boyle said that the next best alternative is a retrospective health study for which the sampling frame is available and the intervention has already taken place. He described the design of a multi-community-based retrospective cohort study in which choosing schools was the first stage of sample selection, followed by selection of particular grades. The approach would be to select schools at a given time, identify a cohort of students who attended the selected schools, and then locate and collect information on the selected cohort. One of the challenges would be proactively identifying prospective study participants, as administrative records may be sealed for various reasons.

Boyle noted the example of the retrospective twin registry study. The Vietnam-era twin registry was a resource of the U.S. Department of Veteran Affairs, composed of 7,369 middle-aged male-male twin pairs, both of whom served in the military during the time of the Vietnam conflict (1965-1975). It was used for genetic epidemiologic studies of mental and physical health conditions. Boyle urged workshop participants to consider twin registries as a vehicle to collect information on criminal justice involvement and said he thinks that they are a great resource for studying counterfactual cases. He concluded his talk by stressing the importance of reducing/minimizing nonresponse bias, cognitive testing of criminal justice questions, and framing the questions in a double-anchored fashion to get accurate responses.

## THE NATIONAL LONGITUDINAL SURVEY OF YOUTH

Elizabeth Cooksey (Ohio State University) discussed measures of criminal justice involvement collected on the 1979 National Longitudinal Survey of Youth (NLSY79), a nationally representative sample of 12,686 young men and women who were aged 14-21 when they were first surveyed in 1979. They were interviewed annually through 1994 and are currently interviewed on a biennial basis. The response rate in the latest round of the survey stands at 78 percent.

Cooksey said the NLSY79 is focused primarily on labor market involvement, but the instrument does have a few criminal justice and health questions. Variables on criminal justice include the earliest arrest date, total number of arrests, earliest incarceration date, total number of incarcerations, age at first incarceration, length of first incarceration, length of longest incarceration, total months incarcerated and current incarceration. The study does attempt to locate and contact survey respondents who are serving time. She suggested that three subsamples of the survey—a supplemental sample designed to oversample minority groups; the NLSY79 child sample consisting of children born to female respondents; and the NLSY79 young adults sample consisting of these same children aged 15 and older—would be of primary interest to the workshop participants. The first sample is useful for understanding the prevalence of criminal justice involvement and health outcomes in minority groups, and the child and young adult samples can provide information on collateral consequences of parental incarceration. Cooksey told the workshop participants that the NLSY97 included about 750 individuals who reported being incarcerated at some point in their lives. The NLSY79 and NLSY97 have an extended health module for respondents of a certain age that contains information on mental and physical health, health conditions, and health care coverage. Beginning in 2006, NLSY79 respondents aged 48 and 49 were given a battery of exercises designed to capture cognitive capabilities.

## THE PANEL STUDY OF INCOME DYNAMICS

David Johnson (University of Michigan) focused on health data collected on the Panel Survey of Income Dynamics (PSID), as well as linkage opportunities. The PSID is a household longitudinal survey that started in 1968 with 5,000 families. The unique feature of the survey is that it covers households that developed from the original households—the family units started by the children of the original 5,000 families. Consequently, there are data available on approximately 1,200 three-generation families, making it the longest running household longitudinal survey in the world. Collecting the survey's detailed health data and biological specimens requires consent from survey respondents and is an expensive process.

The PSID consists of a core interview, which collects self-reported health, health care, and economic data, as well as rotating supplementary modules, focusing on other social and demographic content. The survey gathered information on household members' jail or prison experiences from 1985 to 1992. PSID also conducted a special Childhood Retrospective Circumstance Survey (PSID-CRCS) in 2014 funded by the National Institute on Aging in which they asked participants aged 19 and older about incarceration experience before age 26. Of the over 8,000 responses received, Johnson said about 188 adults in the sample reported that they had been in jail at least once before age 26 (Institute for Social Research, n.d.-b). In another PSID supplement conducted in 2007 called the Transition to Adulthood Study, 254 of the approximately 1,260 18-year-olds surveyed reported that they had been to jail (Institute for Social Research, n.d.-a). Johnson said that, while the data can prove insightful, the sample sizes for the PSID core and supplementary modules often provide challenges when trying to generate valid estimates of health outcomes. Johnson explained that PSID data can be linked to other sources, such as Medicare claims, Housing and Urban Development records, census data, and real estate data. It would be possible to link PSID data to criminal justice and inmate surveys; the challenge is getting consent from survey respondents regarding linking their data to other sources.

## THE NATIONAL CRIME VICTIMIZATION SURVEY

Jim Lynch (University of Maryland), former head of the Bureau of Justice Statistics (BJS), discussed the challenges in capturing a relatively rare phenomenon in household surveys. Referring to the surveys described in this workshop session, he said that most of them do not have a large enough sample size to generate a worthwhile subsample of people who are incarcerated. The traditional approach has been to study these types of phenomena through a small longitudinal or repeated cross-section survey. Often, however, the lack of relevant responses results in expanding the definition of dependent variable (from "did you" statements to "have you ever" statements) in an effort to obtain a sufficient number of observations to conduct an empirical analysis. Consequently, Lynch said, the National Crime Victimization Survey (NCVS) is a useful tool to study criminal justice involvement because of its large sample size of 90,000 households. Even if the single-year sample for incarceration is small, larger samples can be achieved by combining sample sizes across years. The disadvantage of the NCVS is that it does not collect data on health outcomes or health care access or utilization.

Lynch pointed out, however, that there are attributes of the NCVS that make it a good vehicle for fielding questions on criminal justice involvement. The NCVS has a rotating panel design in which a household is interviewed twice in 6 months, making certain to capture every member of the household between those two interviews (or using proxy interviews if that is the only option). Because the survey asks about people who have entered or exited the household, one can sometimes crudely deduce which households have been touched by incarceration (although "moved to an institution" could mean a transition to a nursing home or other non-penal setting). The NCVS does offer the opportunity for externally funded supplements, which could be a potential vehicle for more specific criminal justice involvement questions.

Lynch next described the administrative data linkage initiatives undertaken by BJS and the Census Bureau that can improve the amount of information available on criminal justice involvement and health outcomes. The Census Bureau started the Center for Administrative Record Research and Application (CARRA) to harness the potential of administrative data from federal, state, and third-party providers through record linkage and statistical matching while ensuring confidentiality. He said that CARRA is a terrific model because participating data providers can use linked datasets and analyze data linkages from sources in the criminal justice system.

BJS's correctional unit has undertaken efforts to link data from the National Prisoner Statistics Program and the National Corrections Reporting Program to other data sources, including those with health data. He relayed the experience of BJS in using Social Security numbers as the identifier variable: they are generally a good linking variable for the nonprison population but not for those who have a criminal history record. With that limitation, the agency has been successful in matching certain prison and corrections data cases and observations to administrative data on health. Other strategies tested for multiple years when the Social Security numbers were not available had relatively high success rates. Another potential source of data that allows investigation and differentiation of correctional experience, as well as an expansion into health data, is the National Inmate Survey. BJS is currently in the process of securing consent from inmates so that their data can be linked to other sources: he noted that some inmates have refused to give consent.

Lynch stressed that linking survey data with administrative data is fraught with challenges as there are errors in both data sources, and definitional differences can hamper connecting similar pieces of information. Moreover, he said, accessing administrative data and securing permission to match them with other sources requires a good relationship with the data provider, be it federal, state, or some other third party. Such initiatives are undertaken on a case-by-case basis, and the full potential of administrative data has not been realized.

Lynch advised the workshop participants to strongly consider investigating the linking opportunities that are available in the Federal Bureau of Investigation's Computerized Criminal History (CCH) data system. The CCH data have two major strengths—information on arrests and information on corrections. The utility of criminal history data has not been explored for the purposes of studying health effects of incarceration; it has been mostly used to study recidivism. Lynch concluded his presentation by stressing the importance of building a statistical infrastructure of record linkage that is easily accessible to researchers.

## DISCUSSION

To begin the discussion, Laub invited comments and questions from the presenters in earlier sessions. Candace Kruttschnitt (University of Toronto) said she was struck by the lack of coverage of incarcerated populations in general and social surveys. Marcie Cynamon (National Center for Health Statistics) said that national household health surveys could cover institutionalized populations if their mandates were modified and their budgets considerably expanded in order to meet the objective of understanding health outcomes in a relatively rare and small population. She also noted that the population is very heterogeneous, which will require a larger than normal sample size to conduct empirical analysis for subgroups of interest.

Allen Beck (Bureau of Justice Statistics) informed the workshop participants that the health data captured on inmate surveys conducted by the agency are collected through self-reporting. The data cover information on inmates' health conditions prior to their being incarcerated and the medical conditions they developed while in prison. Beck noted that mortality rates of prison inmates are lower than comparable general non-institutionalized populations (except for liver mortality and sepsis mortality), which he said can be attributed to access to health care services in prisons.

Cynamon raised the issue of the reference period being restricted to only 1 year: limiting a household survey question to events that transpired in the previous year ignores many people who had contact with the criminal justice system early in their lives or at any point before the past year and the effects of that contact. Laub responded by saying that restricting the reference period to 1 year provides an anchor to the respondent that helps in recall, but there is the danger of telescoping, bringing forward the events that happened much earlier. On this point, Anjani Chandra (National Center for Health Statistics) talked about the purpose and focus of surveys: some, like the National Survey of Family Growth, are aimed at producing annual aggregate statistics, and therefore a reference period of 1 year is justified. Life trajectories can be deduced from lifetime experience questions that ask respondents about when they first experienced a particular event. Boyle responded to the point about life trajectories by pointing out that there had been no mention of covariates associated with traumatic experiences, such as resilience and coping. He said this implies that adding a couple of criminal justice questions on a national health survey will not provide that type of information.

Johnson suggested that the American Community Survey (ACS) could be another data source on the health outcomes of the incarcerated population because the instrument contains six questions on disability and is administered to institutionalized populations. There is precedent for using the sampling frame of the ACS for a follow-on survey.

Emily Wang (Yale University) said she surmises that household surveys set national priorities from the standpoint of government and the general public. The reason that incarcerated populations are not on their radar is due to a lack of understanding about the collateral consequences of incarceration among the general public. She pushed for inclusion of measures on criminal justice involvement in national health surveys as the value of such information can further the understanding of the effects of incarceration and criminal justice contact on individuals, families, and communities.

# 8

# Key Themes and Possible Next Steps

John Laub (University of Maryland), Wendy Manning (Bowling Green State University), and Emily Wang (Yale University) led a discussion about key themes presented over the course of the workshop. Wang spoke on how the workshop presentations highlighted some key tensions that arise when considering health equity for incarcerated populations in comparison with non-incarcerated populations. Although there is quite a bit of information about the individual effects of incarceration on health, information on how it affects families, communities, and the population is lacking.

Another outstanding issue, Wang said, is the need for a greater understanding of causality, mechanisms, risk factors, and how criminal justice ultimately affects the life course of an individual. Wang said that as a researcher she understands the desire to pursue the topic with a very narrow and focused approach; however, there is no single set of questions that will address every possible risk factor, and, as such, a single, broad question might be a better place to start. She also pointed out a few of the limitations one may face considering the addition of a criminal justice involvement question to population health surveys: in addition to the scarcity of survey space, the sample sizes of the major health surveys in question do not yield high percentages of people with criminal justice histories.

Wang said she believes there seem to be short-, medium- and long-term goals. The short-term idea would be to add a single question, harmonized across the health surveys, that would provide some understanding of the concept of criminal justice being an enduring trait: she said she thinks a question about conviction would best serve this purpose. For medium- and long-term goals, researchers should consider adjusting the sampling frame to better capture applicable respondents, incorporating additional questions to gain more insight, and seriously consider linking the survey responses with administrative data. Wang also feels that the discourse needs to be broadened beyond the workshop attendees, in an attempt to engage more policy makers, more public health professionals, and more of the general public.

Laub noted several issues that were not discussed much if at all during the workshop, including privatization of prisons, which could affect the quality of care provided, and childhood disadvantage, which manifests itself in adult health. He proposed that the best approach is to maintain a human-centered focus and a sense of urgency to better understand the needs of the relevant population. From a research quality perspective, Laub said careful consideration should be given to the question of whether self-reporting is going to provide the data accuracy desired. He reiterated statements by Ingrid Binswanger (Kaiser Permanente Colorado and University of Colorado School of Medicine) from earlier in the workshop that prison should not be used as a last resort treatment option for individuals best treated in other settings. Lastly, Laub encouraged the workshop participants to try to reimagine the criminal justice system such that it could, in fact, be a positive turning point in the life course of individuals as opposed to a negative turning point.

Manning agreed that the best starting point might be developing one targeted sociodemographic question that could be asked on population health surveys. She said that there is greater utility in using more than one survey vehicle to obtain these data. She opined that criminal justice involvement should be asked about in the same manner as other sociodemographic characteristics, such as race and educational attainment—topics that provide insight into traits that could potentially shape a respondent's experiences. Manning acknowledged that the responses may differ across surveys and geographic region, however, because of varied demographic distribution.

Manning agreed with the need to learn more about causal inferences and said that in order to do so, more individual analyses are needed and researchers need to be mindful of comparison/control groups. Instead of simply looking at incarcerated and non-incarcerated individuals, it may behoove researchers to find individuals who have never been incarcerated but have reported doing activities very similar to ones for which others were arrested and convicted. Manning also made the suggestion that the participants consider the strong linkages between incarceration and drug use for future analyses.

Michael Massoglia (University of Wisconsin) advised the participants not to automatically think of causality and prevalence as mutually exclusive phenomena. He said there could be opportunities to leverage data in a way that would address causality and also shed light on prevalence issues. Massoglia also expressed concern with Wang's suggestion to use conviction as a starting point for a single, widely applied question, since many people who are convicted of a crime never go to prison. He questioned whether getting to the heart of the matter in a single question would require more specificity. Wang responded, saying that the "denominator" question considered for this purpose, whatever it would be, should have a high percentage point and enduring utility as an indicator of risk.

Christopher Wildeman (Cornell University) and Ruth Peterson (Ohio State University, emerita) continued the discussion about the most effective focal point for criminal justice involvement data collection. Wildeman stated that he feels incarceration is most important to pinpoint, but that if conviction is the point of contact to be measured, the focus should be on felony and not misdemeanor conviction. Felony convictions can more significantly alter the course of a person's life because people are required to disclose felony convictions on applications for employment and other services. Peterson agreed that felony conviction is a more important data point than misdemeanor conviction, but asserted that incarceration should receive highest priority if the criminal justice involvement question set is limited to one or two items. Ross Matsueda (University of Washington) assented to the idea of asking about incarceration, with questions about felony conviction being a close second priority if space allowed. Amanda Geller (New York University) suggested using the term "custody" instead of "incarceration," since there are people in jail who have not yet been convicted of a crime.

Anjani Chandra (National Survey of Family Growth) asked the participants if they believed that the time frame of recall was of any importance when formulating a question. Wildeman suggested the question should ask about involvement at any point—a "have you ever" question—rather than a fixed time frame. John Boyle (ICFI International) reminded the group that the item does not have to be a "yes/no" question; it is possible to list the various points of contact as options in one question, and allow for multiple responses.

Michael Schober (The New School) said that he was intrigued by Manning's proposal to incorporate a criminal justice involvement question in a list of demographic questions. He mentioned that, historically, the initial cluster of demographic questions becomes normalized over time. Adding criminal justice involvement to that familiar repertoire could potentially remove some of the stigma associated with the topic. He added that obtaining a better understanding of the relative sizes and sampling frames of the surveys in question could drive the decision making for what type of single question would be best asked. He said this would be particularly helpful when thinking about future policy discussion. Schober also reminded the workshop participants that, to the extent that people want to make any proposal for an existing survey or to any organization that has an ongoing survey that has its own mandate, its own cycle, and the like, researchers need to be very clear in their argument for how adding new data collection items is in the interest of that specific survey.

Binswanger said that information about the facility from which the person receives care should also be taken into account. People who have had criminal justice contact but are still living in the community have access to a different set of services than people in prisons or jails. Binswanger said that in addition to distinguishing the type of care provided, knowledge of the entity responsible for providing the care would facilitate quality assessments.

Rashida Dorsey (U.S. Department of Health and Human Services) thanked the workshop participants for all their input. She then asked the participants to continue contemplating which criminal justice contact points were important to measure in relation to health: to consider them in terms of priority, and to think about what each would help us better understand as well as what it would not be able to tell us. Dorsey encouraged participants to share their later thoughts with the Office of the Assistant Secretary for Planning and Evaluation and the Office of Minority Health in an effort to continue the discourse on this issue.

# References

Aquilino, W.S., Wright, D.L., and Supple, A.J. (2000). Response effects due to bystander presence in CASI and paper-and-pencil surveys of drug use and alcohol use. *Substance Use & Misuse, 35*(6-8), 845-867.

Barton, A.H. (1958). Asking the embarrassing question. *Public Opinion Quarterly, 22*(1), 67-68.

Beckett, K., and Western, B. (2001). Governing social marginality welfare, incarceration, and the transformation of state policy. *Punishment & Society*, *3*(1), 43-59.

Berman, A. (2015). *Give Us the Ballot: The Modern Struggle for Voting Rights in America.* New York: MacMillan.

Biderman, A.D., Cantor, D., and Reiss, A. (1985). *A Quasi Experimental Analysis of Personal Victimization by Household Respondents in the NCS.* Paper presented at the annual meetings of the American Statistical Association, Philadelphia, PA.

Brown, S.L., and Manning, W.A. (2013). *Experiences with the Criminal Justice System in a Household Survey: Introducing the Survey of Criminal Justice Experience (SCJE).* Bowling Green, OH: Bowling Green State University, National Center for Family & Marriage Research. Available: http://ncfmr.bgsu.edu/pdf/Measuring%20Incarceration%20HH%20Surveys/file135154.pdf [December 2016].

Cantor, D. (2016). *Designing Questions on Criminal Justice Involvement for Health Surveys.* Presentation to the National Research Council Workshop on Improving Collection of Indicators of Criminal Justice System Involvement in Population Health Data Programs. Available: http://sites.nationalacademies.org/cs/groups/dbassesite/documents/webpage/dbasse_171858.pdf [December 2016].

Carson, E.A. (2015) *Prisoners in 2014.* Bulletin, Bureau of Justice Statistics, NCJ 248955. Available: http://www.bjs.gov/content/pub/pdf/p14.pdf [November 2016].

Comfort, M. (2008). *Doing Time Together: Love and Family in the Shadow of the Prison.* Chicago, IL: University of Chicago Press.

Corkrey, R., and Parkinson, L. (2002). A comparison of four computer-based telephone interviewing methods: Getting answers to sensitive questions. *Behavior Research Methods, Instruments, & Computers, 34*(3), 354-363.

Durose, M.R., Cooper A.D., and Snyder, H.N. (2014). *Recidivism of Prisoners Released in 30 states in 2005: Patterns from 2005 to 2010.* Washington, DC: U.S. Department of Justice. Available: http://www.bjs.gov/content/pub/pdf/rprts05p0510.pdf [January 2017].

Foster, H., and Hagan, J. (2007). Incarceration and intergenerational social exclusion. *Social Problems, 54*(4), 399-433.

Foster, H., and Hagan, J. (2013). Maternal and paternal imprisonment in the stress process. *Social Science Research, 42*(3), 650-669.

Foster, H., and Hagan, J. (2015). Maternal and paternal imprisonment and children's social exclusion in young adulthood. *Journal of Criminal Law & Criminology, 105*, 387.

Foster, H., and Hagan, J. (2016). *Criminal Justice Involvement & Its Impact on the Individual, Family, and Community.* Presentation to the National Research Council Workshop on Improving Collection of Indicators of Criminal Justice System Involvement in Population Health Data Programs. Available: http://sites.nationalacademies.org/cs/groups/dbassesite/documents/webpage/dbasse_171861.pdf [December 2016].

Geller, A., Jaeger, K., and Pace, G.T. (2016). Surveys, records, and the study of incarceration in families. *Annals of the American Academy of Political and Social Science, 665*, 22-43.

Glaze, L.E., and Maruschak, L.M. (2008). *Parents in Prison and Their Minor Children.* Special Report, Bureau of Justice Statistics, NCJ 222984. Available: https://www.bjs.gov/content/pub/pdf/pptmc.pdf [December 2016].

Guerino, P., Harrison, P.M., and Sabol, W.J. (2012). *Prisoners in 2010.* Washington, DC: Bureau of Justice Statistics.

Hagan, J., and Foster, H. (2012a). Children of the American prison generation: The paradoxical "spillover" school effects of incarcerating mothers. *Law & Society Review, 46*(1), 37-69.

Hagan, J., and Foster, H. (2012b). Intergenerational school effects of mass imprisonment in America. *Sociology of Education, 85*(3), 259-286.

Hagan, J., and Foster, H. (2015). From mass incarceration to the Great Recession: A human security perspective on parental imprisonment and young adult economic insecurity in America. *Russell Sage Journal of the Social Sciences, 1*(2).

Hindelang, M.J., and McDermott, M.J. (1981). *Juvenile Criminal Behavior: An Analysis of Rates and Victim Characteristics.* Washington, DC: U.S. Department of Justice, Law Enforcement Assistance Administration, Office of Juvenile Justice and Delinquency Prevention.

Holtgraves, T., Eck, J., and Lasky, B. (1997). Face management, question wording, and social desirability. *Journal of Applied Social Psychology, 27*(18), 1650-1671.

Huebner, B.M., and Gustafson, R. (2007). The effect of maternal incarceration on adult offspring involvement in the criminal justice system. *Journal of Criminal Justice, 35*(3), 283-296.

Institute for Social Research. (n.d.-a). *The Child Development Supplement Transition into Adulthood Study 2007: User Guide.* Available: https://psidonline.isr.umich.edu/CDS/TA07_UserGuide.pdf [November 2016].

Institute for Social Research. (n.d.-b). *The Panel Study of Income Dynamics' Childhood Retrospective Circumstances Study (PSID-CRCS): User Guide.* Available: https://psidonline.isr.umich.edu/CRCS/2014UserGuide.pdf [November 2016].

Kaeble, D., Glaze, L., Tsoutis, A., and Minton, T. (2015). *Correctional Populations in the United States, 2014.* Bulletin, Bureau of Justice Statistics, NCJ 249513. Available: http://www.bjs.gov/content/pub/pdf/cpus14.pdf [November 2016].

Lee, R.D., Fang, X., and Luo, F. (2013). The impact of parental incarceration on the physical and mental health of young adults. *Pediatrics, 131*(4), e1188-e1195.

Lee, H., McCormick, T., Hicken, M., and Wildeman, C. (2015). Racial inequalities in connectedness to imprisoned individuals in the United States. *Du Bois Review, 12*(2), 269-282.

Lind, L.H., Schober, M.F., Conrad, F.G., and Reichert, H. (2013). Why do survey respondents disclose more when computers ask the questions? *Public Opinion Quarterly, 77*(4), 888-935.

Muftic, L.R., Bouffard, L.A., and Armstrong, G.S. (2016). Impact of maternal incarceration on the criminal justice involvement of adult offspring: A research note. *Journal of Research in Crime & Delinquency, 53*(1), 93-111.

Mumola, C.J. (2000). *Incarcerated Parents and Their Children.* Special Report, Bureau of Justice Statistics, NCJ 182335. Washington, DC: U.S. Department of Justice.

Murray, J., and Farrington, D.P. (2005). Parental imprisonment: Effects on boys' antisocial behaviour and delinquency through the life-course. *Journal of Child Psychology and Psychiatry, 46*(12), 1269-1278.

National Research Council. (2014). *The Growth of Incarceration in the United States: Exploring Causes and Consequences.* J. Travis and B. Western (Eds.). Division of Behavioral and Social Sciences and Education, Committee on Law and Justice, Committee on Causes and Consequences of High Rates of Incarceration. Washington, DC: The National Academies Press.

Preisendörfer, P., and Wolter, F. (2014). Who is telling the truth? A validation study on determinants of response behavior in surveys. *Public Opinion Quarterly, 78*(1), 126-146.

Roettger, M.E., and Swisher, R.R. (2011). Associations of fathers' history of incarceration with sons' delinquency and arrest among black, white, and Hispanic males in the United States. *Criminology, 49*(4), 1109-1147.

Schnittker, J., Massoglia, M., and Uggen, C. (2012). Out and down: Incarceration and psychiatric disorders. *Journal of Health and Social Behavior, 53*(4), 448-464.

Schober, M.F., Conrad, F.G., Antoun, C., Ehlen, P., and Fail, S. (2015). Precision and disclosure in text and voice interviews on smartphones. *PLoS ONE, 10*(6), e0128337.

Sudman, S., and Bradburn, N.M. (1982). *Asking Questions: A Practical Guide to Questionnaire Design.* San Francisco, CA: Jossey-Bass.

Swann, C.A., and Sylvester, M.S. (2006). The foster care crisis: What caused caseloads to grow. *Demography, 43*(2), 309-335.

Tourangeau, R., and Smith, T.W. (1996). Asking sensitive questions: The impact of data collection mode, question format, and question context. *Public Opinion Quarterly, 60*(2), 275-304.

Tourangeau, R., Groves, R.M., and Redline, C.D. (2010). Sensitive topics and reluctant respondents demonstrating a link between nonresponse bias and measurement error. *Public Opinion Quarterly*, *74* (3), 413-432.

Tourangeau, R., Kreuter, F., and Eckman, S. (2015). Motivated misreporting: Shaping answers to reduce survey burden. In U. Engel (Ed.), *Survey Measurements: Techniques, Data Quality, and Sources of Error* (pp. 24-41). Chicago, IL: University of Chicago Press.

Turner, A.G. (1981). The San Jose recall study. In R.G. Lehnen and W.G. Skogan (Eds.), *The National Crime Survey: Working Papers.* NCJ-75374, Bureau of Justice Statistics (pp. 22-27). Washington, DC: U.S. Department of Justice.

Turney, K. (2014). The consequences of paternal incarceration for maternal neglect and harsh parenting. *Social Forces, 92*(4), 1607-1636.

Turney, K., and Haskins, A.R. (2014). Falling behind? Children's early grade retention after paternal incarceration. *Sociology of Education*, *87*(4), 241-258.

Turney, K., and Wildeman. C. (2015). Detrimental for some? Heterogeneous effects of maternal incarceration on child wellbeing. *Criminology & Public Policy, 14*(1), 125-156.

Turney, K., Wildeman, C., and Schnittker, J. (2012). As fathers and felons. *Journal of Health and Social Behavior, 53*(4), 465-481.

Villarroel, M.A., Turner, C.F., Eggleston, E., Al-Tayyib, A., Rogers, S.M., Roman, A.M., Cooley, P.C., and Gordek, H. (2006). Same-gender sex in the United States: Impact of T-ACASI on prevalence estimates. *Public Opinion Quarterly, 70*(2), 166-196.

Wakefield, S., and Wildeman, C. (2011). Mass imprisonment and racial disparities in childhood behavioral problems. *Criminology & Public Policy, 10*(3), 793-817.

Wildeman, C. (2010). Paternal incarceration and children's physically aggressive behaviors: Evidence from the Fragile Families and Child Wellbeing Study. *Social Forces, 89*(1), 285-310.

Wildeman, C. (2014). Parental incarceration, child homelessness, and the invisible consequences of mass imprisonment. *ANNALS of the American Academy of Political and Social Science, 651*, 74-96.

Wildeman, C. (2016). *What Do We Want to Measure When We Measure Criminal Justice Contact? Some Thoughts.* Presentation to the National Research Council Workshop on Improving Collection of Indicators of Criminal Justice System Involvement in Population Health Data Programs. Available: http://sites.nationalacademies.org/cs/groups/dbassesite/documents/webpage/dbasse_171868.pdf [December 2016].

Wildeman, C., and Turney, K. (2014). Positive, negative, or null? The effects of maternal incarceration on children's behavioral problems. *Demography 51*(3), 1041-1068.

Wildeman, C., and Wakefield, S. (2014). The long arm of the law: The concentration of incarceration in families in the era of mass incarceration. *Journal of Gender, Race, and Justice, 17*(2), 367-389.

Yan, T. (2016). *Asking Sensitive Questions in Surveys.* Presentation to the National Research Council Workshop on Improving Collection of Indicators of Criminal Justice System Involvement in Population Health Data Programs. Available: http://sites.nationalacademies.org/cs/groups/dbassesite/documents/webpage/dbasse_171869.pdf [December 2016].

Yan, T., Curtin, R., and Jans, M. (2010). Trends in income nonresponse over two decades. *Journal of Official Statistics*, 26(1), 145.

# Appendix A

# Workshop Agenda

**WORKSHOP ON IMPROVING COLLECTION OF INDICATORS OF CRIMINAL JUSTICE SYSTEM INVOLVEMENT IN POPULATION HEALTH DATA PROGRAMS**

March 29–30, 2016
Keck Center of the National Academies of Sciences, Engineering, and Medicine
500 Fifth Street, NW, Washington, DC

**Tuesday, March 29, 2016**
Open Session, 9:00am–5:00pm
Room 101

**Setting the Stage**

9:00am Call to Order
*Breakfast available in the meeting room*

*Welcome and Introduction*
Brian Harris-Kojetin, *CNSTAT Deputy Director*
J. Nadine Gracia, *OMH Director and Deputy Assistant Secretary for Minority Health*

*Purpose of the Workshop*
Wendy Manning, *Committee Chair*

***Value of the Committee's Work to Criminal Justice Populations***
Glenn Martin, *JusticeLeadershipUSA*

**Understanding What Should Be Measured**

10:00am ***Linkages between Incarceration and Health: Current State of Knowledge***
Michael Massoglia, *University of Wisconsin*

*Exploring connections between criminal justice involvement and various components of health, including drug and alcohol abuse. Considering measurement challenges and the need for changes in data collection, as well as an agenda for new research.*

10:45am Coffee Break

11:00am **Unpacking the term "Criminal Justice Involvement"**

***Defining Criminal Justice Involvement through Various Constructs***
Chris Wildeman, *Cornell University (committee member)*

*A brief overview of existing work/literature that attempts to define criminal justice involvement in terms of demographics such as racial disparity, gender, age, and poverty. Identifying key differences in the mechanisms underlying various states of incarceration and supervision. Defining criminal justice involvement by nature of contact, as well as by length, recency, and prevalence of criminal justice involvement.*

***A First Attempt: BGSU Measuring Incarceration in Household Surveys Study***
Wendy Manning, *Bowling Green State University*

*Lessons learned from the 2012 ASPE-sponsored study, as well as a look at the resulting product, the Survey of Criminal Justice Experience.*

12:30pm Lunch

1:30pm **Direct vs. Indirect Effects of Criminal Justice Involvement**

***Criminal Justice Involvement and Its Impacts on the Individual, the Family, and the Community***
John Hagan, *Northwestern University;* Holly Foster, *Texas A&M University*

*How an individual's criminal justice involvement can affect the individual, his or her family, extended family, and the community at large in many different ways.*

***Health Effects as Consequences***
Christopher Uggen, *University of Minnesota*

*Collateral consequences on health stemming from criminal justice system involvement.*

3:00pm Coffee Break

3:15pm **What More Do We Need to Understand?—A Facilitated Session**
Facilitator—Ross Matsueda, *University of Washington*
Participants—Evelyn Patterson, *Vanderbilt University*; Daniel Nagin, *Carnegie Mellon University*; Ingrid Binswanger, *University of Colorado Medical School*

*Participants will consider the causal components of criminal justice involvement on health; identify the critical unanswered research questions and the data needed to explore those answers (including effective sample sizes); and discuss whether various proposals for data collection would be sufficient for the most critical end uses.*

4:45pm Day 1 Wrap-Up

5:00pm Adjourn

**Wednesday, March 30, 2016**
Open Session, 9:00am–3:00pm
Room 101

**Identifying Best Approaches to Measurement**

9:00am *Breakfast available in the meeting room*
**Collecting Indicators of Criminal Justice Involvement on Existing Health Surveys: Five Core Health Surveys as Potential Vehicles**

Marcie Cynamon, *NCHS (NHIS)*; Wayne Giles, *CDC (BRFSS) (via phone)*; Arthur Hughes, *SAMHSA (NSDUH)*; Anjani Chandra, *NCHS (NSFG)*; Kathryn Porter, *NCHS (NHANES)*

*We will hear about each of these five core datasets—the nature of the survey, the target population, and the purpose of data collection. We will also consider how questions about criminal justice involvement and administrative record linkage could be incorporated within a consent form such as the one used for the NHANES.*

10:30am Coffee Break

10:45am **Insights from Alternative Approaches**
Facilitator—John Laub, *University of Maryland (committee member)*
Panel Members—James Lynch, University of Maryland; Elizabeth Cooksey, *Ohio State University (NLSY)*; John Boyle, *ICF International*; Amanda Geller, *New York University (Fragile Families)*; David Johnson, *University of Michigan (PSID)*

*Experience in collecting information about criminal justice involvement in various scenarios may be helpful in the successful development of linkages to health characteristics. The panel's discussion will center on the following: linking criminal justice involvement to social surveys (such as NLSY, PSID, Fragile Families, Add Health); criminal justice and inmate surveys; and the potential use of administrative records to identify past incarceration.*

12:00pm Lunch

1:00pm **Crafting Potential Questions**

***Asking Sensitive Questions in Surveys***
Ting Yan, *Westat*

***Capturing Criminal Justice Involvement in a Few Simple Questions***
David Cantor, *Westat*

*Presenting strawman examples of potential survey questions. Discussion of strengths/weaknesses with regard to adding questions to differently administered surveys (modality), including potential impact on response rates.*

2:00pm Coffee Break

2:15pm **Wrap-Up Discussion**
Wendy Manning, *Bowling Green State University*; John Laub, *University of Maryland*; Emily Wang, *Yale University*

*Committee members' summaries and impressions of the workshop.*

3:00pm Adjourn

# Appendix B

# Biographical Sketches of Steering Committee Members and Speakers

INGRID BINSWANGER *(Speaker)* is a senior investigator at the Institute for Health Research at Kaiser Permanente Colorado. She is also an associate professor in the Division of General Internal Medicine at the University of Colorado School of Medicine, where she directs the primary care research fellowship and the Patient-Centered Outcomes Research Training Program. Her research focuses on preventing the medical complications of substance use, such as overdoses, in high-risk populations. She also conducts research on the health of people involved in the criminal justice system and the risks associated with care transitions between prison and community. She has served as a visiting fellow to the Bureau of Justice Statistics, where she helped develop a framework for the collection of national health indicators on criminal justice populations. She has an M.S. in health and medical sciences from the University of California, an M.A. in public health from the University of Washington, and an M.D. from the University of California, San Francisco.

JOHN BOYLE *(Speaker)* is senior vice president and leads the survey research practice at ICF International. Previously, he held senior positions at Abt SRBI, and Louis Harris and Associates and was on the graduate teaching faculty of the University of Maryland and the research faculty of Columbia University's School of Public Health. His work focuses on the design, execution, analysis, and reporting of large-scale health surveys. He has also conducted surveys of intimate partner violence; rape; other forms of sexual assault and stalking; crime victims' experience with police, prosecutors, courts, parole boards, and other components of the criminal justice system; and the first survey of the prevalence of posttraumatic stress disorder among U.S. ground troops in Vietnam. He has directed health surveys for many federal agencies, including the Centers for Disease Control and Prevention, the National Institutes of Health, the Food and Drug Administration, the Office of Technology Assessment, the Department of Veterans Affairs, and the Department of Defense.

**DAVID CANTOR** ***(Speaker)*** is a senior statistical fellow at Westat and a research professor at the Joint Program for Survey Methodology at the University of Maryland. His work is in designing social surveys and evaluating the effects of social policy covering a range of criminological and methodological topics, including the correlates of victimization risk, the measurement of personal victimization, survey nonresponse and nonresponse bias, and the use of incentives on surveys and panel conditioning. He is currently conducting research on the design of victimization surveys with a particular focus on the collection of data on sexual violence. He has a Ph.D. in sociology from the University of Illinois at Urbana-Champaign.

**ANJANI CHANDRA** ***(Speaker)*** is a demographer/health scientist in the Reproductive Statistics Branch of the Division of Vital Statistics of the National Center for Health Statistics. She conducts research on the fertility and reproductive health of women, including pregnancies and birth, infertility, surgical sterilization, and related aspects of union and family formation. Her primary research interests—social and family demography, reproductive and perinatal health, and survey research methods—are represented in numerous publications and presentations. In her work with the National Survey of Family Growth, Dr. Chandra designs survey instruments that reflect current data needs of the demographic and health research community; tests and evaluates questionnaires; oversees interviewer training, fieldwork, and data collection; and prepares the data for public use, including data cleaning, variable construction, and documentation. She has a B.S. in biology from Yale University and a Ph.D. in demography with a minor in epidemiology from the Johns Hopkins University School of Public Health.

**ELIZABETH COOKSEY** ***(Speaker)*** is professor of sociology at Ohio State University. Her main areas of research focus on social demography, life-course transitions, and the development of youth and children. She is currently studying adolescent sexual and contraceptive behaviors, effects of parents' lives on children, religious beliefs and behaviors of youth, and demographic transitions among the Amish. She is also principal investigator for the Young Adult Survey of the National Longitudinal Study of Youth. Dr. Cooksey has a B.A. in human sciences from Oxford University and an M.A. and a Ph.D. in sociology from Brown University.

**MARCIE CYNAMON** *(Speaker)* is director of the Division of Health Interview Statistics at the National Center for Health Statistics of the Centers for Disease Control and Prevention. The Division plans and executes the National Health Interview Survey and related projects. Her primary area of expertise is health survey research methods.

**HOLLY FOSTER** *(Speaker)* is an associate professor in the Department of Sociology at Texas A&M University and a current visiting scholar at the American Bar Foundation. Her main areas of research are on parental incarceration effects on children, children's exposure to violence, and women's imprisonment. She co-organized the 2013 White House conference on parental incarceration in the United States, sponsored by the National Science Foundation. She has published research on the effects of parental incarceration in the *Annual Review of Sociology*, *Law & Society Review*, *Social Problems*, *Sociology of Education*, *Social Science Research*, *The Annals of the American Academy of Political and Social Sciences*, and *The Russell Sage Journal of the Social Sciences*.

**AMANDA GELLER** *(Speaker)* is a clinical associate professor of sociology at New York University, where she is also an affiliate of the Population Center and director of the MA Program in Applied Quantitative Research. Her research examines the interactions between criminal justice policy and practice, public health, and socioeconomic disadvantage and their joint effects on urban neighborhoods, families, and individuals. She focuses primarily on the administration of justice related to police–public interactions and on the role of incarceration in urban families. She also works on methodological issues for data collection related to both incarceration and police–public interactions. She has a B.S. in operations research and industrial engineering and an M.S. in engineering from Cornell University, and a Ph.D. in social policy analysis from Columbia University.

**WAYNE GILES** *(Speaker)* is director of the Division of Population Health at the National Center for Chronic Disease Prevention and Health Promotion at the Centers for Disease Control and Prevention. His work has included studies examining the prevalence of hypertension in Africa, clinical trials evaluating the effectiveness of cholesterol-lowering agents, and studies examining racial differences in the incidence of stroke. He currently directs research and programmatic activities in a number of areas, including arthritis, aging, alcohol, chronic obstructive pulmonary disease, prevention research, school health, and epilepsy. He has been the recipient of numerous awards, including the Centers for Disease Control and Prevention's Charles C. Shepard Award in Assessment and Epidemiology and the Jeffrey P. Koplan Award. He has a B.A. in biology from Washington University, an

M.S. in epidemiology from the University of Maryland, and an M.D. from Washington University.

**JOHN HAGAN** *(Speaker)* is the John D. MacArthur professor of sociology and law at Northwestern University and codirector of the Center on Law & Globalization at the American Bar Foundation. He is the author of *Who Are the Criminals? The Politics of Crime Policy from the Age of Roosevelt to the Age of Reagan* and *Iraq and the Crimes of Aggressive War.* He co-organized the 2013 White House Conference on parental incarceration in the United States, funded by the National Science Foundation. He is a recipient of the Stockholm Prize in criminology, the Harry Kalven Prize from the Law & Society Association, and the Cesare Beccaria Gold Medal from the German Society of Criminology.

**ART HUGHES** *(Speaker)* is a mathematical statistician at the Center for Behavioral Health Statistics and Quality at the Substance Abuse and Mental Health Services Administration. He has over 30 years of experience in survey design and methodological and epidemiological studies related to substance use and associated behaviors. He is currently overseeing state and substate analysis and reporting activities on the National Survey on Drug Use and Health. Previously, he was chief of the Epidemiology Research Branch at the National Institute on Drug Abuse and a mathematical statistician at the Bureau of Labor Statistics. He has a B.S. in mathematics from Morgan State University and an M.S. in mathematics from Howard University.

**DAVID JOHNSON** *(Speaker)* is deputy director of the Panel Study of Income Dynamics and research professor at the Institute for Social Research at the University of Michigan. He has worked widely throughout the federal statistical system, including his most recent position as chief economist in the Bureau of Economic Analysis. His research focuses on inequality and poverty measurement, equivalence scale estimation, and consumption. Dr. Johnson has a B.S. in economics and mathematics from the University of Puget Sound and a Ph.D. in economics from the University of Minnesota.

**CANDACE KRUTTSCHNITT** *(Member, Steering Committee)* is professor of sociology at the University of Toronto. Previously, she was on the faculty of the University of Minnesota and a visiting fellow at the University of Helsinki, Cambridge University, the Vrije University Amsterdam, and the Netherlands Institute for the Study of Crime and Law Enforcement. She has published widely on women's imprisonment, gender and crime, and victimization. Her current research encompasses work on female offenders and comparative penology. She is a former president of the American Society of

Criminology. Dr. Kruttschnitt has a B.A. in criminology from the University of California–Berkeley and a Ph.D. in sociology from Yale University.

JOHN LAUB *(Member, Steering Committee)* is a distinguished university professor in the Department of Criminology and Criminal Justice at the University of Maryland. Previously, he served as the director of the National Institute of Justice at the Department of Justice—a Senate-confirmed presidential appointment. His areas of research include crime and the life course, crime and public policy, and the history of criminology. He is a fellow of the American Society of Criminology and a past president of the American Society of Criminology. He is a recipient of the Edwin H. Sutherland Award from the American Society of Criminology and (with Robert Sampson) of the Stockholm Prize in criminology research on how and why offenders stop offending. Two books he has co-authored with Robert Sampson—*Crime in the Making: Pathways and Turning Points Through Life*, and *Shared Beginnings, Divergent Lives: Delinquent Boys to Age 70*—have won several major awards. He has a B.A. in criminal justice from the University of Illinois at Chicago and an M.A. and a Ph.D. from the School of Criminal Justice at the State University of New York at Albany.

JAMES LYNCH *(Speaker)* is professor and chair of the Department of Criminology and Criminal Justice at the University of Maryland. Prior to this position, he served as the director of the Bureau of Justice Statistics of the U.S. Department of Justice. Previously, he was a distinguished professor in the Department of Criminal Justice at John Jay College, City University of New York chair, and a professor in the Department of Justice, Law and Society at American University. His research focuses on victim surveys, victimization risk, the role of coercion in social control, and crime statistics. He has published several books and numerous articles dealing with crime statistics. He is president-elect of the American Society of Criminology and served on the Committee on Law and Justice Statistics of the American Statistical Association. Dr. Lynch has a B.A. from Wesleyan University and an M.A. and a Ph.D. in sociology from the University of Chicago.

WENDY MANNING *(Chair, Steering Committee)* is the director of the Center for Family and Demographic Research and codirector for the National Center for Family & Marriage Research (NCFMR). As a family demographer, her research examines how family members define and understand their obligations to each other in an era of increasingly diverse and complex family relationships. She cochaired the 2012 NCMFR Measuring Incarceration in Household Surveys Invitational Forum, which brought together leading scholars, federal data providers, and policy makers to discuss the rise in mass incarceration in the United States, its effects on the well-being of individuals and families, and how to capture and measure the incarceration experience in household surveys. Her work has examined the meaning of cohabitation with her work on the measurement of cohabitation, fertility in cohabiting unions, the stability of cohabiting unions, transitions to marriage, and implications of cohabitation for adult and child well-being. She has a B.A., an M.S., and a Ph.D. in sociology, all from the University of Wisconsin–Madison.

GLENN E. MARTIN *(Speaker)* is the founder of JustLeadershipUSA, an organization that empowers people most affected by incarceration to drive policy reform and is dedicated to cutting the U.S. correctional population in half by 2030. He is a national leader and criminal justice reform advocate who spent 6 years in New York State prisons. Previously, he served as vice president of public affairs at The Fortune Society and as codirector of the National HIRE Network at the Legal Action Center. He is cofounder of the Education from the Inside Out Coalition. He has been an Echoing Green Fellow, an America's Leaders of Change National Urban Fellow, and a member of the governing boards of the College and Community Fellowship, Million Hoodies, and the California Partnership for Safe Communities. He serves on many national boards, including Governor Cuomo's Reentry and Reintegration Council, the advisory board of the Vera Institute's Public Health and Mass Incarceration Initiative, the National Network for Safe Communities, the Executive Session on Community Corrections at Harvard University, and the Global Advisory Council of Cornerstone Capital Group. He was named on the 2015 Root 100 list of most influential African Americans. He regularly contributes his expertise to national news outlets such as MSNBC, Fox News, CNN, Al Jazeera, and CSPAN.

MICHAEL MASSOGLIA *(Speaker)* is a Romnes Professor of Sociology at the University of Wisconsin–Madison, where he is also a Vilas associate of the College of Letters and Sciences and serves as the director of the Center for Law, Society, & Justice. His research focuses on the social consequences of the expansion of the penal system, the relationship between the use of legal controls and demographic change in the United States, and patterns and consequences of criminal behavior over the life course He is an award-winning instructor who teaches classes on criminology, delinquency, and deviance. Dr. Massoglia has a Ph.D. from the University of Minnesota.

ROSS MATSUEDA *(Speaker)* is the Blumstein-Jordan endowed professor of sociology at the University of Washington, where he was previously associate chair of the Department of Sociology and associate director of the Center for Statistics and the Social Sciences. He previously held faculty positions at the University of Wisconsin–Madison and at the University of

Iowa, where he was a professor, chair of the Department of Sociology, and founding director of the Center for Criminology and Socio-Legal Studies His research has focused on classical theories of crime, such as differential association, social control, and labeling. He has also examined the implications of rational choice theories and the theories about crime. His current research includes a study of neighborhood social capital and codes of violence, a study of life course trajectories of substance use and crime, a study of rational choice and deterrence (using panel data from the Denver Youth Survey), and a study of collective action and political protest in Seattle and Leizpig, Germany. Dr. Matsueda has a Ph.D. from the University of California–Santa Barbara.

**DANIEL NAGIN** *(Speaker)* is the Teresa and H. John Heinz III university professor of public policy and statistics at Carnegie Mellon University and serves as the Associate Dean of Faculty of the university's Heinz School Public Policy and Management. He previously served as Deputy Secretary for Fiscal Policy and Analysis in the Pennsylvania Department of Revenue. His research focuses on the evolution of criminal and antisocial behaviors over the life course, the deterrent effect of criminal and non-criminal penalties on illegal behaviors, and the development of statistical methods for analyzing longitudinal data. He is an elected fellow of the American Society of Criminology, the American Society for the Advancement of Science, and the American Academy of Political and Social Science. He is a recipient of the American Society of Criminology's Edwin H. Sutherland Award and of the Stockholm Prize in Criminology. He has a Ph.D. from what is now the Heinz School of Public Policy and Management of Carnegie Mellon University.

**EVELYN PATTERSON** *(Speaker)* is assistant professor of sociology at Vanderbilt University. She is also a health policy associate in the Center for Health Policy at Meharry Medical College. She was previously assistant professor of crime, law and justice, sociology, and demography at Pennsylvania State University. She has received awards from the American Sociological Association and the Population Association of America. She has published widely on sentencing policies, health and mortality, and measurement issues in correctional data. Dr. Patterson is the recipient of awards from the American Sociological Association and the Population Association of America. She has an M.A. in demography and a joint Ph.D. in criminology and demography from the University of Pennsylvania.

**RUTH PETERSON** *(Member, Steering Committee)* is professor emerita of sociology at Ohio State University and former director and a faculty retiree pf the university's Criminal Justice Research Center (CJRC). She is the co-organizer of the CJRC's Racial Democracy, Crime, and Justice-Network and its Crime and Justice Summer Research Institute: Broadening Perspectives and Participation and a member of CJRC's Spatial Crime Research Working Group. She currently serves as a member of the Committee on National Statistics and vice-chair of the Committee on Law and Justice at the National Academies of Sciences, Engineering and Medicine. Her research focuses on community conditions and crime, racial and ethnic inequality in patterns of crime, and the consequences of criminal justice policies for racially and ethnically distinct communities. She is currently serving as president of the American Society of Criminology. She has a Ph.D. in sociology from the University of Wisconsin.

**KATHRYN PORTER** ***(Speaker)*** is director of the Division of Health and Nutrition Examination Surveys (DHANES) at the National Center for Health Statistics. In that capacity, she is responsible for managing the planning and implementation of the ongoing National Health and Nutrition Examination Survey program and oversees the overall analytic research activities for the survey. She is also a captain in the U.S. Public Health Service Commissioned Corp and a former officer in the Epidemic Intelligence Service. She serves as a medical officer in the Operations Branch of DHANES. She is a member of the American Medical Association, the American Public Health Association, and the American College of Preventive Medicine. Dr. Porter is board certified in preventive medicine and public health and is a fellow of the American College of Preventive Medicine. She has an M.S. in preventive medicine and epidemiology from the University of Maryland and an M.D. from the Medical College of Virginia.

**JOSIAH RICH** ***(Member, Steering Committee)*** is professor of medicine and epidemiology at the Warren Alpert Medical School of Brown University and attending physician at the Miriam Hospital, where he is also the cofounder and director of the Center for Prisoner Health and Human Rights at the hospital's Immunology Center. His work focuses on infectious diseases and addiction, and he has advocated for public health policy changes to improve the health of people with addiction, including increasing drug treatment for the incarcerated and formerly incarcerated populations. He has published widely on the overlap between infectious diseases, addictions, and incarceration. He is the principal investigator of three major studies focused on incarcerated populations. He is a cofounder of the nationwide Collaboration in HIV in Corrections Initiative of the Centers for AIDS Research. Dr. Rich has advocated for public health policy changes to improve the health of people with addiction, including improving legal access to sterile syringes and increasing drug treatment for the incarcerated and formerly incarcerated populations. He has an M.P.H. from the Harvard School of Public Health and an M.D. from the University of Massachusetts.

MICHAEL SCHOBER *(Member, Steering Committee)* is professor of psychology at the New School for Social Research in New York City. His research deals with questions that cross the lines between psychology, linguistics, human–computer interaction, music, public opinion research, and design. Dr. Schober's recent and ongoing studies include work on conversational language use and perspective-taking, how differently people can conceive of what they are discussing despite apparent understanding, how partners with differing abilities take each other into account, conceptual misalignment in survey interview and testing interactions, how IQ testers can influence responses and scores, how survey interviewing techniques affect response accuracy, how interacting with interviewing systems that are more and less human-like affects survey respondents' willingness to disclose personal information, comprehension of natural speech (including disfluencies and stutters), interface design and interaction, and how attention to respondent disfluencies and other "paradata" can be useful for interviewing interfaces. He has a Ph.D. in psychology from Stanford University.

CHRISTOPHER UGGEN *(Speaker)* is Martindale Chair and Distinguished McKnight Professor of Sociology and Law at the University of Minnesota. He studies crime, law, and deviance. He co-wrote *Locked Out: Felon Disenfranchisement and American Democracy* s and has written extensively on felon voting, work and crime, and harassment and discrimination and is frequently cited in the media. His research, teaching, and advising interests include crime and punishment, social and political inequality, and law and society. Current projects include a comparative study of reentry from different types of institutions, employment discrimination and criminal records, crime and justice after genocide, and the health effects of incarceration. He has a B.A. in criminal justice and an M.S. and a Ph.D. in sociology, all from the University of Wisconsin.

EMILY WANG *(Member, Steering Committee)* is an associate professor at the Yale School of Medicine and co-founder of the Transitions Clinic Network, a national network of medical homes for individuals with chronic diseases recently released from prison. Her research focuses on promoting health equity for vulnerable populations, especially individuals with a history of incarceration, through both prison- and community-based interventions. She works in training former prisoners to become community health workers and in training researchers through community-based participatory research methods. She is cofounder of the Transitions Clinic Network, a consortium of 15 community health centers nationwide dedicated to caring for recently released prisoners and defining best practices for the health care of individuals leaving prison. In 2012 the Transitions Clinic Network was awarded the Centers for Medicare & Medicaid Services Innovation Award to provide care to over 2,000 high-risk, high-cost patients returning from prison and to train and employ former prisoners as community health workers. Dr. Wang has a B.A. from Harvard University, an M.A. from the University of California–San Francisco, and an M.D. from Duke University.

**CHRISTOPHER WILDEMAN** ***(Member, Steering Committee)*** is an associate professor of policy analysis and management in the College of Human Ecology at Cornell University, where he is also a faculty fellow at the Bronfenbrenner Center for Translational Research, the Center for the Study of Inequality, Court-Kay-Bauer Residence Hall, and the Cornell Population Center. He has been a visiting fellow at the Bureau of Justice Statistics. Previously, he was on the faculty at the Center for Research on Inequalities and the Life Course and the Institution for Social and Policy Studies, both at Yale University. He was also codirector of the New Haven Branch of the Scholars Strategy Network. His research and teaching interests revolve around the consequences of mass imprisonment for inequality, with emphasis on families, health, and children. Dr. Wildeman is a recipient of the Ruth Shonle Cavan Young Scholar Award from the American Society of Criminology. He has an M.A. and a Ph.D. in sociology and demography from Princeton University.

**TING YAN** ***(Speaker)*** is a senior survey methodologist at Westat and an adjunct faculty member at the Joint Program in Survey Methodology at the University of Maryland and at the Program in Survey Methodology at the University of Michigan. Her work in the field of survey research focuses on questionnaire design, question evaluation, paradata, and evaluation of data quality. She has a Ph.D. in survey methodology from the Joint Program in Survey Methodology at the University of Maryland.

## COMMITTEE ON NATIONAL STATISTICS

The Committee on National Statistics was established in 1972 at the National Academies of Sciences, Engineering, and Medicine to improve the statistical methods and information on which public policy decisions are based. The committee carries out studies, workshops, and other activities to foster better measures and fuller understanding of the economy, the environment, public health, crime, education, immigration, poverty, welfare, and other public policy issues. It also evaluates ongoing statistical programs and tracks the statistical policy and coordinating activities of the federal government, serving a unique role at the intersection of statistics and public policy. The committee's work is supported by a consortium of federal agencies through a National Science Foundation grant.